Undergraduate Topics in Computer Science

'Undergraduate Topics in Computer Science' (UTiCS) delivers high-quality instructional content for undergraduates studying in all areas of computing and information science. From core foundational and theoretical material to final-year topics and applications, UTiCS books take a fresh, concise, and modern approach and are ideal for self-study or for a one- or two-semester course. The texts are all authored by established experts in their fields, reviewed by an international advisory board, and contain numerous examples and problems, many of which include fully worked solutions.

The UTiCS concept relies on high-quality, concise books in softback format, and generally a maximum of 275–300 pages. For undergraduate textbooks that are likely to be longer, more expository, Springer continues to offer the highly regarded Texts in Computer Science series, to which we refer potential authors.

More information about this series at https://link.springer.com/bookseries/7592

Tobias Weinzierl

Principles of Parallel Scientific Computing

A First Guide to Numerical Concepts and Programming Methods

 Springer

Tobias Weinzierl
Department of Computer Science
Durham University
Durham, UK

ISSN 1863-7310   ISSN 2197-1781   (electronic)
Undergraduate Topics in Computer Science
ISBN 978-3-030-76193-6   ISBN 978-3-030-76194-3   (eBook)
https://doi.org/10.1007/978-3-030-76194-3

This Springer imprint is published by the registered company Springer Nature Switzerland AG
The registered company address is: Gewerbestrasse 11, 6330 Cham, Switzerland

Preface

Why This Book

I started to think about this manuscript when I got appointed as Associate Professor at Durham University and had been told to read numerical algorithms for computer scientists plus the parallel programming submodule. In Durham, we teach this first submodule on numerics in the third year of the computer science degree. It spans only ten hours. Ten hours is ridiculously short, but I think it is enough to spark interest in simulation codes and the simulation business. Thinking about it, it is great that our numerics teaching runs parallel to a parallel programming course. Numerical simulation without parallelism today is barely imaginable, so this arrangement gives us the great opportunity to use examples from one area in the other one. We can co-teach the two things: numerics basics and parallel coding.

When we later designed our Master in Scientific Computing and Data Analysis (MISCADA), it became clear that co-teaching the two fields would be a good idea for this course, too. With a very heterogeneous student intake—both by means of qualification, cultural background, educational system, and discipline interest—we need to run introductory courses that bring all students quickly to a decent level, while numerical principles as well as parallel programming are omnipresent. Obviously, the M.Sc.-level teaching progresses at a different pace and, in Durham, also covers more lectures. Yet, some content overlaps with the third year's course.

There seems to be a limited amount of literature that discusses parallel programming and (the basics of) numerical techniques in one go and still remains easily accessible. There are many books that cover scientific computing from a mathematics point of view. Many scientists (including me) insist that a formal approach with a strong mathematical background is imperative for research in scientific computing. However, we have to appreciate that many of our undergrads are enrolled in courses that are lighter on the mathematics side or focus, for example, more on discrete maths. Some maths thus might be a little bit rusty when students enter their third year. Our mathematical textbooks run the risk that we lose out on students if they orbit around the profound maths first. They might get bored before they reach the point where they can program a first numerical code that does "something useful".

The other type of teaching literature that is out there combines the first course in numerics with a programming tutorial. Although a lot of the material I have seen is really good, this type of approach is ill-suited for our students. Most of them are decent programmers already.

To make a long story short, I think that there is a need for courses on numerics with a "rolling up your sleeves" and "get the rough idea and make it work in code" attitude. Some students like to get their hands dirty and write something "that does something" before they start to dive into more details—for which excellent literature does exist. I am also strongly convinced that it makes sense to teach introductory scientific computing in combination with parallelisation. After all, we want to have future Ph.D. students and colleagues who can help us with the big simulation codes.

Mission Statement

Since we try to keep the theory light and hands-on, this book is not a comprehensive overview of the vast field of numerical algorithms and analysis, or even only a certain subfield. It neither is a rock solid introduction to the difficulties, pitfalls and depths of programming in high performance computing (HPC). It is

- a rush through some basic numerical concepts that are omnipresent when we write simulation software. We want to get an idea of how some numerical things work—so we can eventually look them up in a "real" maths book.
- a numerics primer with implementation glasses on: For all of the concepts, we want to have an idea of how to realise them on a real machine or which impact they have on our codes.
- a hands-on text. The best way to understand numerical concepts is to program them or, even better, to see your own code fail and then find a theoretical explanation (together with the fix) for the break down. And it is close to impossible to learn parallel programming solely from textbooks anyway.
- a guide that makes students start to write simulation codes running on parallel computers.

Structure and Style

The manuscript jumps from parallel programming to numerics and back. I sprinkle in some light-touch revision of things that I consider to be important but have been told by my students that many computer science curricula do not cover them anymore. Even if taught at school before, students might have forgotten them. Where possible, I try to make all content feed into one simple N-body solver. That's the application leitmotif throughout the manuscript. If readers translate it into a working code, they will end up with an N-body solver with simple gravity, will have learned fundamental ideas behind the solver such as numerical terminology or challenges faced by integrators for ordinary differential equations, and will have the skill set to make this solver run efficiently on a shared memory machine or some GPUs. There's no need for a reader (or teacher) to stick to my concept or to read the chapters one by one. Feel free to skip chapters, pick chapters, or use a different application example.

Mathematicians will be horrified by my hands-on explanations and lack of formalism. I am fine with this. This book is not yet another maths course. It is a book that shall be quick and easy to read and motivate students to study the maths behind the scenes later, once the first generation of their simulation code is up and running. This book is also not a course where students really learn something about N-body codes for a particular application area. Simple N-body codes just seemed to be a convenient toy problem that we can understand with A-levels maths.

Experts deeply familiar with the details of OpenMP, e.g., also will be horrified by the inaccuracies in the text. For example, the visibility of a variable in the OpenMP standard is determined through a sequence of rules. These rules are important, but most students quickly will be lost in them. Also, most programmers don't have to memorise them. They have to know what happens in 99% of the cases.

As I want to push students into the field of scientific computing plus HPC, I made all examples in the text rely on C/C++. I am well aware that many codes still are written in Fortran, and I am well aware that people nowadays can write interesting and fast software with Python, e.g. But C still is very dominant and, fortunately, many computer science students are still exposed to C in one form or the other throughout their undergraduate studies. While A-levels maths is sufficient to digest the book's content, I hence expect students to bring along C programming skills.[1]

Maybe unusual for a student-facing book, this text does not contain any exercises. There are two reasons for this: On the one hand, there are plenty of books, in particular in numerics, that are full of exercises; and more and more webpages with (interactive) examples follow. On the other hand, the prime way to exercising scientific computing is to write codes, and I think that there is a big difference between writing small, well-defined example calculations or a bigger piece of software. With this text, I want to encourage students to do the latter.

Acknowledgements

Thanks are due to numerous students who provided feedback. Particular thanks are due to my colleagues Florian Rupp and Philipp Neumann. They found tons of bugs and helped me to fix a lot of phrases. Thanks are also due to Lawrence Mitchell for proofreading some of the chapters with an HPC touch. Holger Schulz worked with

---

[1]The UTICS series hosts, for example, the Pitt-Francis and Whiteley book "Guide to Scientific Computing in C++" where Chaps. 1–5 provide a reasonable base for the present book.

me on the ExCALIBUR/ExaClaw project throughout the write-up and helped me quite a lot with the OpenMP-focussed chapters. Marion Weinzierl finally was maybe the most critical reader, but eventually it had been her comments and her support over all the years that made writing this manuscript possible.

Learning and Teaching Mode

Major parts of this manuscript have been written during the first Corona lockdown. I am sure that this unexpected situation has forced many lecturers to revise, rethink and re-evaluate their academic teaching style. I had to. Though I do not know (yet) whether this book is of any use or help for the brand new world of blended and virtual teaching, the manuscript style is a result of my personal "rethink your teaching" exercise.

The appendix hosts some ideas on how I use the book's content to structure my classes. The text is written in a language such that students can smoothly work through the content themselves: If I have to make a choice, I run for the sloppy jargon rather than ultra-formal, dryish phrases. Furthermore, I tried to use illustrations for technical details, as I am myself a visual learner. Key learning outcomes are extracted and organised into boxes, and the appendix hosts some checklists and cheatsheets for students' coursework. All of this allows me to use the text as a basis for courses in a "flipped classroom"-style. Obviously, this style can be complemented by online/synchronous sessions, videos, exercises and code studies; and, obviously, reading lists that dive deeper into the subject.

Enjoy the text, and I hope that it is useful.

Durham, UK   Tobias Weinzierl
January 2022

Contents

Wrap-up

Part I

Introduction: Why to Study the Subject

The Pillars of Science

1

Abstract

We discuss the two classic approaches (pillars) to acquire new insight in science: experiment and theory. As computer simulations supersede the classic experiment and blackboard theory, we obtain a third pillar. This notion of a third pillar can be challenged, but it motivates us to define new research areas assembled under the umbrella of the catchphrase computational science and engineering. After a quick sketch of the simulation pipeline—typical steps required in any computational discipline—we introduce scientific computing as area conducting research into the tools enabling insight through computing.

This is a thought experiment, but feel free to make it a real one: We take a heavy object (not grandma's favourite Ming vase) and a light one such as our pen and stand on a chair in our office. We drop both objects. Which one hits the ground first?

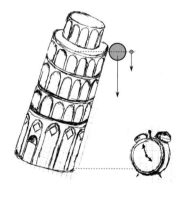

T. Weinzierl, *Principles of Parallel Scientific Computing*, Undergraduate Topics
in Computer Science, https://doi.org/10.1007/978-3-030-76194-3_1

The experiment is a classic in Physics in school. The urban legend says that
Galileo Galilei (1589–1592) stood on the Leaning Tower of Pisa and dropped
objects when he served as professor of mathematics there. The experiment
studies how the mass of an object affects the speed with which it falls down.
We might assume that the heavier object falls down faster. Aristotle had taught
exactly this, so the assumption is actually not that naïve or stupid. Galileo's
experiment shows—as long as we neglect effects such as friction—that the objects' speed does not depend on the mass. It thus invalidates Aristotle's idea:
The "arrival" time of the object (when it crashes into the floor) is independent
of its mass.

We are not interested here in who did and thought what. This is not a history
of science lesson. The story illustrates something else: In science, we often
have a theory and an experiment, and one thing either validates or invalidates
the other one.

Humans and in particular scientists make *models* of the world all the time. We
postulate theories how things work and why. Every discipline has its own language
that help us to create these models. In science and engineering, mathematics is the
modelling language of choice.

With a model at hand, we compare its predictions with reality, i.e. with experiments. Let's assume $T(m)$ is a function that gives us—for our tower experiment—the
time how long an object with mass m needs to hit the ground. The Aristotelian
hypothesis is that the heavier an object the smaller T. In the words of maths:
$m_1 > m_2 \Rightarrow T(m_1) < T(m_2)$. The observations from the experiment teach us
that this is not correct. The experiment teaches us that $T(m)$ is independent of m.

The interplay of theory and experiment is not a one-way street. New models
drive scientists to come up with new experiments to validate them, while the model
development is driven by observations: Whenever we observe something that we
cannot explain with the ideas and theories we have at hand, we ask ourselves "why is
this" and then start to develop a more sophisticated model which captures the things
we just observed.

> The two pillars of science

Theory and experiment are the two classic pillars (or legs) of science. They validate
and push each other. Theoretical scientists and engineers make all kinds of models
of the real world, but they are considered to be valid, i.e. we consider them to be reasonable accurate representations of the real world, only after we have measurements
(data) that follow the predictions of these theories. If the experiments yield different
results or show effects that we can't explain, the theory guys have to go back to the
blackboard.

1.1 A Third Pillar?

The two pillars as we have introduced them so far are barely enough to explain how we advance in science and engineering today.

- *Some experiments are economically infeasible.* If we want to design the next generation of jumbo jets, we might want to put each model into a wind tunnel to see whether it does take off. But fuelling a wind tunnel (in particular on a reasonable scale) is extremely expensive.
- *Some experiments are ethically inappropriate.* If we design novel ways of radiation treatment, it would not be ethical to try this out with patients in a trial-and-error fashion. Another example: If we construct a new bridge, we don't want the first cars driving over this bridge to be Guinea pigs.
- *Some experiments are ecologically dubious.* If we design novel nuclear reactors, we don't want to rely on trial-and-error when it comes to security.
- *Some experiments are by construction impossible.* If we make up new theories about the Big Bang, we are lost with our two pillar model: we will likely never be able to run a small, experimental Big Bang.
- *Some equations are so complex that we cannot solve them (analytically).* For many setups, we have good mathematical models. Maybe, we can even make claims about the existence and properties of solutions to these models. But that does not always mean that the maths gives us a constructive answer, i.e. can tell us what the solution to our model is.

This list is certainly not comprehensive. Its last point is particularly intriguing in modern science: We have some complex equations comprising a term $u(x)$. Let this $u(x)$ be the quantity that we are in interested in. Yet, we cannot transform this formula into something written as $u(x) = \ldots$ with no $u(x)$ on the right-hand side. That is, even when we have x, we still do not know $u(x)$ directly.

A simple example goes as follows: The acceleration of a free object in space is determined by gravity. Acceleration is the second derivative of the object's position in time. The first one is the velocity. With a formula for the gravitational force F, we know $\partial_t \partial_t u(x) = CF$ (where ∂_t is the time derivative). This is something we can solve analytically, i.e. write down the solution—either by hand or via a computer program or a simple websearch. With two free objects that attract each other, solving the arising equations analytically, i.e. distilling $u(x) = \ldots$, becomes tricky. For three, it becomes impossible.

All statements on the shortcomings of theory and experiment illustrate an important trend: More and more areas of science need a computer to compensate for the challenges traditional experiments and theoretic approaches face. Engineers simulate in a computer how a bridge behaves under load before they build it. Physicists plug their new Big Bang models into a Universe simulation, make this simulation compute what the Universe would look like today, and then compare the outcome to observations. Mathematicians ask a computer to plot the solutions to complex equations calibrated to some measurements they know and thus obtain a feeling whether a

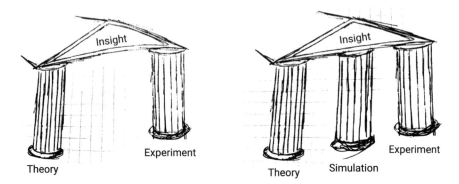

Fig. 1.1 Left: New insight relies on the interplay of theory and experiment in the classic model. Right: As theory and experiment often fall short, simulations become a third pillar of insight. Or not?

model represents reality. More and more disciplines heavily rely on computer simulations. This makes people speak of them as third pillar of insight besides experiment and theory (Fig. 1.1).

However, there are people who challenge this notion of a third pillar. After all, there are few things in nature with three legs (Fig. 1.2). And at the same time, there are people who claim that we even have a fourth pillar today: big data bringing in machine learning and artificial intelligence. The argument against the three pillar concept orbits around the observation that most theoretical work today is not a pen-and-paper exercise anyway. Mathematicians use computers all the time to approximate and solve equations. Experimentalists in return have always used machines to track, bookkeep and interpret their data; starting from a slide ruler or a notebook. The four pillar advocates argue that we live in an age where data explodes and computer algorithms obtain insight (learn) from these data autonomously; without complex models of what they are supposed to see in the data.

I am a big fan of the two-pillar view of the world. It is not only about keeping things simple, it is about re-reading the two classic legs of science, theory and experiments,

Fig. 1.2 My "insight" ape never runs on three legs. It either stands on two of them supplemented by data processing (in particular machine learning) and simulation, or we read data and simulation as two additional arms besides the theory and experiment leg

through today's glasses. Read Moshe Y. Vardi's discussion of why science has only two legs (see below on further reading). He points out that we can think of science as theory and modelling plus experiments and analysis of data. Computer simulations are always used whenever we come up with a new theory or model. And when we do experiments, we are typically swamped with data. Without computers and notably simulations which give us a clue what to search for, it is hard to find the right thing in all of our measurements. Artificial intelligence fits in here, too—as "yet" another tool to digest vast experimental data.

Definition 1.1 (*The third pillar of science*) Some people call computer simulations the *third pillar of science*. Others disagree. Anyway, computer simulations are indispensable in modern science—whether as pillar of its own or as essential part of theory and experiment.

1.2 Computational X

We call the disciplines X that rely on computer simulations Computational X. There's Computational Engineering, Computational Physics, Computational Chemistry, Computational Biology, and so forth. We often use Computational Science and Engineering (CSE or CS&E) as an umbrella term covering all the different flavours. Today, this is a little bit of an old-fashioned phrase, since we don't want to exclude Computational Medicine, Computational Finances and so forth. They all share similar challenges. Lacking a better term, let's stick to CSE.

A typical project in Computational X brings together expertise and skills from three traditional areas: the application discipline, Mathematics and Computer Science (Fig. 1.3). The term CSE thus covers a broad church of challenges:

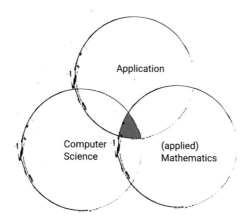

Fig. 1.3 The three areas coming together in Computational X projects. These projects employ (generic) tools and knowledge from Scientific Computing. In return, they stimulate the involved disciplines

1. The modelling of physics or application knowledge with mathematical equations;
2. the transcription of these equations into something (other equations) that a computer can solve. Often that means breaking down an infinite fine (continuous) model into a finite number of equations;
3. the analysis of these equations: do they have a solution, how reliable is an (approximate) solution, and so forth;
4. the design of an algorithms solving these equations;
5. the analysis of these algorithms: how expensive will it be to run them, i.e. what's the algorithmic complexity, e.g.;
6. the coding of these algorithms;
7. the systematic testing of these codes;
8. the performance optimisation to make the codes run fast and scale on big machines;
9. the input and output data management;
10. the postprocessing ranging from visualisation to pattern searches within outputs;
11. …

This sequence and variants thereof are called the *simulation pipeline*. I give a first, simple example of the most prominent steps within this pipeline in Chap. 3. I personally dislike the term pipeline. It suggests—similar to the term waterfall in software development—some kind of sequentiality. In practice, we jump around and even make excursions into the theory and experiment world. Let's turn our attention to the generic challenges within CSE and within this pipeline:

> Chris Johnson: "Before every great invention was the discovery of a new tool".[1]

This is a statement that holds in many different ways: Brunelleschi not only invented the double shell system when he constructed the cupola (dome) of the Duomo of Florence, he also first constructed some novel cranes that allowed him to get the roof on. Reinforced concrete first had to be invented before mankind had been able to construct some large bridges. There are many such examples. We need (to invent, discover luckily, or pile up and assemble the set of) the right tools to solve problems. If we want to solve problems on a computer, we need the right computational tools, i.e. mathematics, algorithms, data structures, programming techniques, …We need the right tools for Computational X. This book covers some aspects of these tools.

> Scientific Computing

We work within the realm of Scientific Computing whenever we work on the tools that eventually allow Computational X to make progress.

---

[1] From an interview in the video "The Golden Age of Computing".

While the challenges in Scientific Computing derive from particular Computational X challenges, many of Scientific Computing's tools, algorithms and ideas are generic. They are not tied to a particular problem. This makes it an area of research of its own.

1.3   About the Style of the Text and Some Shortcomings

In this manuscript, we look at Scientific Computing with Computer Science glasses on. Within the simulation pipeline, we have a certain bias. Other Scientific Computing books start from stronger mathematical grounds or stick to a particular application domain. We focus on getting some software up and running, on understanding some underlying principles, and on ideas how to make the software run efficiently.

Further reading

- David E. Keyes: *Computational Science*. In N.J. Higham: The Princeton Companion to Applied Mathematics. pp. 335–350 (2015)
- Moshe Y. Vardi: *Science Has Only Two Legs*. Communications of the ACM. 53(9), page 5 (2010) http://doi.acm.org/10.1145/1810891.1810892
- Jaenette M. Wing: *Computational Thinking*. Communications of the ACM. 49 (3), pp. 33–35 (2006) https://doi.org/10.1145/1118178.1118215
- Michael Sharratt. *Galileo: Decisive Innovator*. Cambridge University Press (1994)

> **Key points and lessons learned**

- We speak of three pillars of science: experiment, theory and simulation. Some doubt that three-pillar notion.
- A discipline X which relies on simulations for new insight, is also called Computational X.
- Scientific computing studies the tools (algorithms, software, data structures, …) that enable progress in the computational disciplines.
- We have a rough idea of a generic simulation pipeline, i.e. tasks we have to complete to solve a problem in Computational X.

Moore Myths

2

Abstract

We introduce Moore's Law and Dennard scaling. They illustrate that further increases of compute capabilities will result from an increase in parallelism over
the upcoming years; which implies that our codes have to be prepared to exploit
this parallelism. We identify three levels of parallelism within a CPU (inter-node,
intra-node and vector parallelism) and eventually characterise GPUs within these
levels.

In the early Middle Ages, the construction business was guided by experience.
A simple rule went like that: The larger a building such as a church, the higher
its weight. Higher weight means more massive walls to carry the weight. You
also should only have small windows, so you literally get a rock-solid structure.

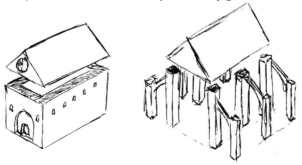

With the advent of the Gothic architecture came the insight that it might be
cleverer to use a stone rib structure that is filled out with light walls, vaults
or even glass windows. The rib structure carries the weight and distributes it
properly. Larger, bigger, more impressive buildings became feasible.

© The Author(s), under exclusive license to Springer Nature Switzerland AG 2021   11
T. Weinzierl, *Principles of Parallel Scientific Computing*, Undergraduate Topics
in Computer Science, https://doi.org/10.1007/978-3-030-76194-3_2

> The transition of building styles is a paradigm change: We can either have four walls and make them bigger and fatter and, hence, more robust. Or we can use tiny walls (actually ones that are degenerated into columns). Each individual one is small and weakish, but there are a lot of them—collaborating to carry our roof.

The expressiveness of simulations hinges on the availability of compute power. It is intuitively clear that a computer that we use for simulations is the "more useful" the faster it yields results. Scientists don't have to wait for results that long or can run more simulations per time. They get more results, and they get them quicker. Often, this is not how it works: We give scientists and engineers a faster computer, and they immediately run a more complex experiment. So we either speed up our scientific workflow, or we help to increase the expressiveness of our calculations.

2.1 Moore's Law

For decades, computational scientists have been in a comfortable situation: They wrote code and made this code fast on a particular architecture. Everybody knew that architectures evolve kind of continuously, i.e. with every new generation of machines "old"ish codes ran faster, too. They might not benefit from the latest hardware features, but there was a performance improvement. Some people claim this were Moore's law, which is not correct. Let's revisit this "law"[1] :

> ### Moore's scaling

Gordon Moore, one of the co-founders of Intel, observed that the cost to put transistors onto a chip decreases if we squeeze more transistors on the circuit. From a certain point on, however, the manufacturing cost rises again, since the integration of all the transistors becomes expensive. Consequently, there's a sweet spot: a magic number of transistors per chip where the chip is most profitable. Moore observed that the "complexity for minimum component costs has increased at a rate of roughly a factor of two per year". So the number of transistors on a chip around the sweet spot grows exponentially according to this law. The manufacturing sweet spot moves and therefore vendor designs move with the spot.

Intel's executive David House later corrected the statement—to 18 months—so it is even more aggressive, while Carver Mead from CalTech coined the term "Moore's Law". Today, the law continues to hold though the rate of the increase has slowed down (Fig. 2.1).

---

[1] The "law" is so popular that Intel paid 10,000 USD for a copy of the original paper print in 2005.

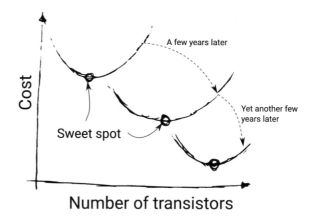

Fig. 2.1 Illustration of Moore's law: Over time, vendors have been able to integrate more and more transistors into a single chip given a fixed budget envelope

For simulation codes as we have sketched them before, it is not directly clear why the transistor count makes a difference. We are interested in speed. However, there is a correlation: First, vendors use the opportunity to have more transistors to allow the computer to do more powerful things. A computer architecture provides some services (certain types of calculations). With more transistors, we can offer more of these calculation types, i.e. broaden the service set. Furthermore, vendors use the opportunity to squeeze more cores onto the chip. Finally, the more of transistors historically did go hand in hand with a shrinkage of the transistors. This leads us to another important law:

2.2 Dennard Scaling

Definition 2.1 (*Dennard scaling*) The power cost P to drive a transistor follows roughly

$$P = \alpha \cdot CFV^2.$$

This law is called *Dennard scaling*.

C is the capacitance, i.e. encodes the size of the individual transistors. I use C here for historic reasons. In the remainder of this manuscript, C is some generic constant without a particular meaning. F is the frequency and V the voltage. α is some fixed constant so we can safely skip it in our follow-up discussions. Note that the original Dennard scaling ignores that we also have some leakage. Leakage did not play a major role when the law was formulated in 1974.

Dennard's scaling law is all about power. For both chip designers and computing centres buying and running chips, controlling the power envelope of a chip is a sine qua non, as

- buying power is expensive, and as
- a chip "converts" power into heat. To get the heat out of the system again requires even more power to drive fans, pumps and cooling liquids. But if we don't get it out of the system on time, the chip will eventually melt down.

While we want to bring the power needs down, we still want a computer to be as capable as possible. That means, it should be able to do as many calculations per seconds as possible. The Dennard scaling tells us that we have only three degrees of freedom:

1. Reduce the voltage. This is clearly the gold solution as the V term enters the equation squared. Reducing the voltage however is not trivial: If we reduce it too much, the transistors don't switch reliably anymore. As long as we need a reliable chip, i.e. a chip that always gives us the right answer, we work already close to the minimum voltage limit with modern architectures.
2. Reduce the transistor size. Chip vendors always try to decrease transistor sizes with the launch of most new chip factories or assembly lines. Unfortunately, this option now is, more or less, maxed out. You can't go below a few atoms. A further shrinkage of transistors means that the reliability of the machine starts to suffer—we ultimately might have to add additional transistors to handle the errors which once more need energy. Most importantly, smaller chips are more expensive to build (if they have to meet high quality constraints) which makes further shrinking less attractive.
3. Reduce the frequency. If we reduce the frequency, we usually also get away with a slightly lower voltage, so this amplifies the savings effect further. However, we want to have a faster transistor, not a slower one!

> **Frequency of computers stagnates or even reduces**

Without altering the frequency, chip vendors struggle to reduce the power needs of a transistor further, while the energy hunger of big computers starts to dominate the computer cost and to outshadow the procurement investments, as the transistor count continues to grow. As a result, vendors freeze or even reduce the chip frequencies.

We asked for faster chips but what we get instead is plateauing or even decreasing frequencies. There used to be a time where roadmaps predicted 10 GHz desktops. This has never happened for the reasons above. For computational scientists, this means that we can't rely on upcoming machine generations to run our codes faster automatically. We have to prepared that our codes become slower unless we manage to exploit the increasing number of transistors.

2.3   The Three Layers of Parallelism

> **A paradigm shift**

There used to be a time where the increase of transistors materialised primarily in new compute features and, at the same time, it paired up with higher and higher frequencies. Nowadays, we still see an increase of the transistor count, but the frequency (speed of individual transistors) stagnates or is even reduced.

With Moore's law continuing to hold and a break-down of Dennard scaling (frequency cannot be increased anymore at a given power envelope per transistor), there is a "new" kid on the block that helps us to build more powerful computers. This one eventually allows Computational X to run more challenging simulations. Actually, there are three new kids around that dominate code development today (Fig. 2.2). However, they all are flavours of one common pattern:

1. The parallelism in the computer increases, as modern computers still can do one addition or multiplication or …on one or two pieces of data in one (abstract) step.[2] They apply with the same operation to a whole vector of entries in one rush. We call this *vector parallelism*.
2. The parallelism in the computer increases, as modern CPUs do not only host one core but an ensemble of cores per chip. Since this multicore idea implies that all cores share their memory we call this *shared memory parallelism*.
3. The parallelism in the computer increases, as modern modern supercomputers consist of thousands of compute nodes. A compute nodes is our term for a classic computer which "speaks" to other computers via a network. Since the individual nodes are independent computers, they do not share their memory. Each one has memory of its own. We therefore call this *distributed memory parallelism*.

The three levels of parallelism have different potential to speed up calculations. This potential depends on the character of the underlying calculations as well as on the hardware, while the boundaries in-between the parallelism flavours are often blurred.

   We discuss ways to program two out of three levels of parallelism here, but skip the distributed memory world that unfolds its full power in multi-node supercomputers. The role of accelerators (GPGPUs) in such a classification spans all three rubrics: An accelerator is a computer within the computer including its own memory. GPGPUs themselves rely heavily on both vectorisation and multicore processing: They host multiple (streaming) processors, and the individual processors can handle one instruction acting on multiple pieces of data per step. The latter realisation is traditionally significantly different to what we find in CPUs—this lockstepping

---

[2] Some of these operations might require multiple cycles when expressing them by means of chip frequency, but the punchline is that there's one operation per cycle sequence.

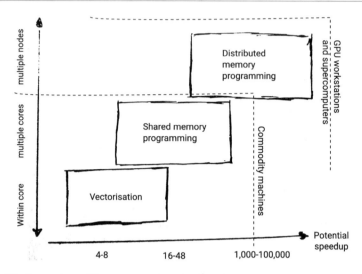

Fig. 2.2 We discuss three different types of parallelisation techniques in this book: Vectorisation, shared memory and distributed memory parallelisation. The potential gains in performance on the x-axis (the labels there) are typical quantities from a classical 2020 CPU-only machine. They will change dramatically over the next few years. Yet, the classification in principle will remain valid

difference is subject of discussion later—even though this distinction starts to blur with new GPGPU generations. Despite the fact that GPUs have a memory that is separated from the CPU, recent programming models succeed in hiding this data exchange; hardware supports this trend. To the programmer, the CPU and the GPGPU start to seem to share memory and they can thus be programmed as a shared memory machine.

> **Three levels of parallelism drive the increase of computer speed**

There are three predominant dimensions of hardware parallelism (vectorisation, multicores and multinodes) along which computers gain performance, while the individual scalar building blocks stagnate in speed.

We have to write code, right from the start, that accommodates the different types of parallelism and that can exploit increasing hardware parallelism in the future. If a new machine is installed after a year's time, only codes that are flexible w.r.t. parallelism can benefit from it. This is not easy. Herb Sutter has likely summarised it best: "The free lunch is over". Programmers have to be skilled in different parallel programming techniques, and they have to think right from the start about how they write future-proof codes w.r.t. parallelism. Parallelisation is not a nice gimmick or add-on that you start to study once you got the functional behaviour right. In this manuscript, I therefore introduce parallelisation techniques hand in hand with the numerical techniques.

Some interesting (historical) material

- Herb Sutter: *The Free Lunch Is Over—A Fundamental Turn Toward Concurrency in Software*. Dr. Dobb's Journal. 30(3) (2005) http://www.gotw.ca/publications/concurrency-ddj.htm
- Gordon E. Moore: *Cramming more components onto integrated circuits*. Electronics Magazine. 38(8), pp. 33–35 (1965) https://doi.org/10.1109/N-SSC.2006.4785860
- Robert H. Dennard et al.: *Design of ion-implanted MOSFET's with very small physical dimensions*. IEEE Journal of Solid State Circuits. 9(5), pp. 256–268 (1974) https://doi.org/10.1109/JSSC.1974.1050511

> **Key points and lessons learned**

- Moore's law is all about transistor counts, i.e. complexity per cost. This used to translate 1:1 into performance growth for years. It does not anymore.
- Dennard scaling states that the power consumption of a chip scales linearly with the frequency but quadratic with the voltage. Its historic formulation ignores leakage which is becoming a non-negligible factor in modern chips.
- If we can't reduce the voltage anymore, we can not increase the frequency as the ship shrinkage is naturally constrained. Vendors have to increase the parallelism on the chip.
- Codes and algorithms have to be redesigned and programmed right from the start to fit to massively parallel systems.
- We distinguish three different levels/types of parallelism on modern hardware in this book.
- GPGPUs are technically separated from the computer. Modern hardware and programming techniques hide this fact and integrate GPGPUs better and better. Logically, we can thus treat them as shared memory and vector computing devices.

Our Model Problem (Our First Encounter with the Explicit Euler)

3

Abstract

A simple N-body problem serves as demonstrator for the topics discussed in upcoming sections. We introduce an explicit Euler for this showcase problem informally, and thus obtain a baseline toy code to play with. The introduction allows us to discriminate numerical approximations from analytical solutions.

We return to our Leaning Tower thought experiment and remind ourselves that the force acting on an object is $F = mg$ with $g = 9.81 \frac{m}{s^2}$ being a magic constant[1]. m in $F = mg$ is the mass of the object that we drop. We also know from school that the acceleration of an object is the derivative of its velocity, i.e.

$$\frac{\partial}{\partial t}v(t) = a(t) \quad \text{while } F = ma.$$

We integrate and plug in our knowledge to obtain

$$v(t) = \int_0^t a(\tilde{t})d\tilde{t} = \int_0^t g \, d\tilde{t}.$$

Here, we've exploited the fact that the object that we drop is at rest when we release it up there on the tower. If the dropped object had an initial velocity, we would have to add a constant term $v(0)$ to the integral above.

Next, we remind ourselves that velocity is the change over time of the position $p(t)$. When we drop an object from $p(0) > 0$ over the ground, we can compute the object's position through

---

[1] The notation here uses an m in $\frac{m}{s^2}$ as unit, while the m in $F = mg$ denotes a parameter. From hereon, I work quite lazily with units and just skip them.

© The Author(s), under exclusive license to Springer Nature Switzerland AG 2021
T. Weinzierl, *Principles of Parallel Scientific Computing*, Undergraduate Topics in Computer Science, https://doi.org/10.1007/978-3-030-76194-3_3

$$p(t) = p(0) - \int_0^t v(\tilde{t})\, d\tilde{t} = p(0) - \int_0^t \int_0^t g\, d\tilde{\tilde{t}} d\tilde{t} = p(0) - g \cdot \frac{1}{2} \tilde{t}^2 \Big|_{\tilde{t}=0}^{t}.$$

Therefore, we can compute exactly when the object crashes into the ground, i.e. when $0 = p(t) = p(0) - g\left(\frac{1}{2}t^2 - \frac{1}{2}0^2\right) = p(0) - \frac{9.81}{2}t^2$.

This was easy. And it was wrong. In real life, the object that we drop and the Earth attract each other. And the force thus follows the equation $F = G \cdot \frac{m_1 \cdot m_2}{r^2}$ with a constant $G = 6.67408 \cdot 10^{-11}$. $F = mg$ is an approximation. It is reasonably accurate, as our dropped object is tiny compared to Earth and so far away from its centre. Therefore, we can neglect the change of the position of the Earth, and we can assume that the distance to the Earth's centre does effectively not alter throughout the free fall.

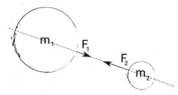

For a massive object (or two planets in space), we have to work with the real formula. This time, things become ugly as the antiderivative is not trivial anymore:

$$v_n(t) = \int_0^t a(\tilde{t}) d\tilde{t} = \int_0^t F(\tilde{t})/m_n\, d\tilde{t} = \int_0^t G \frac{m_1 \cdot m_2}{m_n \cdot r^2(\tilde{t})}\, d\tilde{t}.$$

Let's sit on object 1 and observe how far object 2 is away. That is, we make object 1 rest at position 0, while the other one moves freely. We move the coordinate system with p_1. Then,

$$p_2(t) = p_2(t = 0) - \int_0^t \int_0^t \frac{m_1 \cdot m_2}{m_2 \cdot r^2(\tilde{t})}\, d\tilde{\tilde{t}} d\tilde{t} = p_n(0) - \int_0^t \int_0^t \frac{m_1}{p_2^2(\tilde{t})}\, d\tilde{\tilde{t}} d\tilde{t}.$$

With a sophisticated computer program, some search in the Internet or a proper maths book you'll find a solution, but it is not something anymore that everybody can write down straightaway.

Let's finally use three objects rather than only two (Earth, Moon and the Sun, e.g.). Can you compute a solution and determine a collision point? Likely not. Is there even a distinct collision point, or do the three objects flatter around each other?

For many problems in science and engineering, we have good models that describe how individual parts of the system behave. We have, for example, formulae that describe how rigid bodies (planets) move under gravity. Unfortunately, that does not imply that we can always solve the equations that arise once we use the relatively simplistic equations as building blocks within a larger setup. The N-body problem as sketched above (for $N \in \{2, 3\}$) is a classical example for a relatively simple problem, i.e. the underlying mechanics are simple equations. However, we cannot rewrite the physical model into a formula alike $p(t) = \ldots$ which gives us the position p of the objects for a given time t directly once $N \geq 3$. To be fair, there's a sum representation for $N = 3$, but even that one is irrelevant for practical applications as it is infinitely long.

3.1 The N-Body Problem

Given are N bodies (planets, e.g.) out there in space. They interact with each other through gravity. Let body 1 be described by its position $p_1 = (p_{x,1}, p_{y,1}, p_{z,1})^T \in \mathbb{R}^3$ (a vector) and its mass m_1. Furthermore, it has a certain velocity $v_1 = (v_{x,1}, v_{y,1}, v_{z,1})^T \in \mathbb{R}^3$ which is a vector again. Body 2 is defined analogously. Body 1 experiences a force

$$F_1 = G \frac{m_1 \cdot m_2}{|p_1 - p_2|_2^2} \cdot \frac{(p_2 - p_1)}{|p_1 - p_2|_2}. \tag{3.1}$$

If there are more than two objects, then Body 1 also gets a contribution from Body 3, 4, and so forth. The forces simply sum up. We furthermore know that

$$\partial_t v_1(t) = \frac{F_1}{m_1} \quad \text{and} \tag{3.2}$$
$$\partial_t p_1(t) = v_1(t). \tag{3.3}$$

This is a complete mathematical model of the reality. It highlights in (3.2) and (3.3) that velocity and position of our object depend on time. These equations are our whole theory of how the world out there in space behaves (cmp. Chap. 1) in a Newton sense. Einstein has later revised this model.[2]

Notation remark

The expressions $\partial_t y(t) = \frac{\partial}{\partial t} y(t)$ both denote the derivation of a function $y(t)$. Often, we drop the (t) parameter—we have already done so for F above. For the second derivative, there are various notations that all mean the same: $\partial_t \partial_t y = \partial_{tt} y = \frac{\partial \partial}{\partial t \partial t} y$. I often use $\partial_t^{(2)}$. This notation makes it easy to specify arbitrary high derivatives.

---

[2] Einstein developed his theories, since the classic Newton model struggled to explain certain effects: The misfit of Newton's theory with observations—Newton's theory cannot quantiatively correctly predict how light "bends" around objects, though it was pointed out in 1784 (unpublished) and 1801 that it does bend to some degree—was well-known since the 1859s, but it took till 1916 until someone came up with a new theory that explains the effect. After that, experimentalists needed another three years plus a total solar eclipse to run further experiments which eventually validated Einstein's theories. This is a great historic illustration of the ping-pong between the two classic pillars of science.

Besides ∂_t, I also use d_t. This is in line with a lot of literature in mathematics and physics. There is no difference between the two of them as long as we deal with a plain function $f(t)$ only. However, if we have an $f(t, x(t))$ with two arguments where both arguments depend on t—one of them is the t, the other one accepts t as argument—then ∂_t is the derivative where we alter the direct t argument only. ∂_x or $\partial_{x(t)}$ is the derivative w.r.t. the second variable. They both are *partial derivatives*. They look "in one direction".

d_t in contrast it the total derivative w.r.t. t. If you revise your chain rule and the derivatives with multiple arguments, you recognise

$$d_t f(t, x(t)) = \partial_t f(t, x(t)) + \partial_{x(t)} f(t, x(t)) \cdot \partial_t x(t).$$

The ∂_t wobbles around with t but leaves $x(t)$ invariant. In the second summand, the t remains constant, but we vary $x(t)$.

I try to use ∂_t where possible. So whenever I switch to d_t, you may assume that a function enters the equation that has multiple direct or indirect t-dependencies, and that you have to be extra careful and use your knowledge about derivatives of several variables,

---

The force in (3.1) is the bigger the bigger the product of the masses of the two objects. It decreases rapidly if the objects are far away from each other. The Eukledian

$$|p_1 - p_2|_2 = \sqrt{\left(p_{x,1} - p_{x,2}\right)^2 + \left(p_{y,1} - p_{y,2}\right)^2 + \left(p_{z,1} - p_{z,2}\right)^2}$$

denotes the distance of two object from each other. It is the r (distance) we have used before in the introductory example,

$$p_2 - p_1 = \begin{pmatrix} p_{x,2} - p_{x,1} \\ p_{y,2} - p_{y,1} \\ p_{z,2} - p_{z,1} \end{pmatrix}$$

is a direction vector pointing from object 1 to object 2. It gives the whole force a direction. Assume we had a force $\hat{F}$ and we computed its directional force as $\hat{F}(p_2 - p_1)$. This would yield a force with a direction. However, the further away the two bodies from each other the bigger this vector, i.e. the bigger its magnitude. This is clearly not what we want, so we have to normalise the direction vector. We have to make it have unit length. This is achieved by the division through $|p_1 - p_2|_2$. We end up with a normalisation of r^{-3} overall. One of the rs is kicked out again via the direction vector.

We next study the last equation. The ∂_t in (3.3) is the time derivative. The faster the position of an object changes over time, the higher its velocity. If the velocity equals zero, the object does not move. The change of the velocity in (3.3) is something we call acceleration. If we apply a force to an object, this force accelerates the object. This acceleration is the more significant the lighter the object. I can push my bike but I struggle to push my car, even if I always accelerate them with my maximum force. We immediately see that our model stands the test of Galileo's experiments from Sect. 1. The gravity force's scaling with the body mass and the acceleration scaling cancel each other out.

Definition 3.1 (*Differential equation*) An equation with a function y where the function's (time) derivatives enter the equation, too, describes a *(system of) differential equations*. So $y(t) = t^2$ is a normal (explicit) equation, but $\partial_t y(t) = t^2$ is a differential equation. A normal equation is brilliant for scientists: once we fix t, we directly know what $y(t)$ is. Differential equations can become difficult or even impossible to rewrite into an normal, explicit formula such that we can evaluate the outcome directly.

We call our differential equations ordinary, i.e. ordinary differential equation (ODE), as there are only derivatives w.r.t. one unknown (time). We formalise the term ODE later (Chap. 12).

3.2 Time Discretisation: Running a Movie

With given positions $p_n(t = 0)$, $i \in \{1, 2, \ldots, N\}$, we can throw all of our analysis skills onto the Eqs. (3.1)–(3.3), and hope that we can find an expression for $p_n(t)$. This is laborious. Bad news: Without further assumption it is impossible to solve this problem.

Movie universe

Let's assume that we live in a movie universe where the whole universe runs at 60 frames per second. In each frame, we know all objects p_n as well as their velocities v_n. We can compute the forces at this very frame. They push the particles, i.e. they alter their velocities. Due to the (changed) velocity, all objects end up at a different place in the next frame (Fig. 3.1). The longer the time in-between two frames, the more the particles can change their position (the velocity acts longer). To make the impact of our force pushes realistic, the forces are the bigger the bigger the time

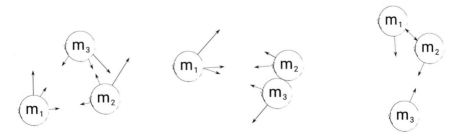

Fig. 3.1 Three frames (left to right) of our movie universe. Forces act in-between the particles but they only update the velocity once per frame. Therefore, we might miss collisions, and we are definitely not exactly mirroring the behaviour of our "continuous" Universe

in-between two frames, too. The more frames per second we use, the more accurate our universe. With the time step size going to zero, we eventually end up with the real thing.

The construction of the movie universe from the real one is called *discretisation*. We know that time is continuous—at least at our scale—and computers struggle to hold things that are infinitely small and detailed. After all, we have finite memory, finite compute power and finite compute time. So we *sample* or *discretise* the time and track the solution in *time steps*. Let dt be the *time step size*, i.e. the time in-between two snapshots.[3] Our computer program simulation the space bodies then reads as follows.

```
// only the declaration here
void calcForce(
  double pAx, pAy, pAz, // position body A [in]
  double pBx, pBy, pBz, // position body B [in]
  double& Fx, double& Fy, double& Fz // force [out]
);


// time stepping loop
...
double t = 0.0;
while (t < T) {                    // we assume T is defined
  for (int n=0; n<N; n++) {    // for every body in system
    // reset forces
    force[n][0] = 0.0; force[n][1] = 0.0; force[n][2] = 0.0;
    for (int m=0; m<N; m++) { // compute interactions
      if (n!=m) {                 // but only with other objects
        double Gx, Gy, Gz;
        calcForce( p[n][0], p[n][1], p[n][2],
          p[m][0], p[m][1], p[m][2], Gx, Gy, Gz
        );
        force[n][0] += Gx;        // accumulate forces
        force[n][1] += Gy; force[n][2] += Gz;
      }
```

---

[3] The notation here can be quickly misunderstood, so you have to be careful: ∂_t and d_t are derivatives. dt is the time in-between two time steps.

```
    }                                   // forces are there, now we can
  }                                     // update particles (don't mix)
  for (int n=0; n<N; n++) {   // for every body in system
    p[n][0] += dt * v[n][0]; p[n][1] += dt * v[n][1];
    p[n][2] += dt * v[n][2];
    v[n][0] += dt * F[n][0]/m[n];
    v[n][1] += dt * F[n][1]/m[n];
    v[n][2] += dt * F[n][2]/m[n];
  }
```

This code is not particular sophisticated. For example, we do not exploit any symmetries between the particles. It is also not necessary to introduce force variables in this code, as we could directly add the force contributions to the object velocities. I leave it to the reader to come up with more intelligent versions of this code. For the time being, it is sufficient to us to know that the scheme we have just implemented is an *explicit Euler*.

Writing data into a file is very slow. Therefore, programs typically give you the opportunity to specify the time in-between two snapshots. This time is usually way bigger than dt. Don't mix them up: You step through your simulation with dt and then every k of these time steps you write a snapshot. As alternative to time steps, you can specify a second DT after which you write the snapshot.

3.3   Numerical Versus Analytical Solutions

Computers are calculators. If we pass them a certain problem like "here are two bodies interacting through gravity", they yield values as solution: "the bodies end up at position x, y, z". They yield numerical solutions. This is different to quite a lot of maths we do in school. There, we manipulate formulae and compute expressions like $F(a, b) = \int_a^b 4x\,dx = 2b^2 - 2a^2$. Indeed, many teachers save us till the very last minute (or until we have pocket calculators) from inserting actual numbers for b and a.

In programming languages, we often speak of variables. But these variables still contain data at any point of the program run. They are fundamentally different to variables in a mathematical formula which might or might not hold specific values. We conclude: There are two different ways to handle equations: We can try to solve them for arbitrary input, i.e. find expressions like $F(a, b) = 2b^2 - 2a^2$. Once we have such a solution, we can insert different a and b values. The solution is an *analytical solution.*[4] On a computer, we typically work *numerically*. We hand in numbers, and we get answers for these particular numbers. But we do not get any universal solution.

Definition 3.2 (*Analytical versus numerical solution*) If we solve an equation via formula rewrites such as integration rules, we obtain an *analytical solution* over the variables. Analytical solutions describe a generic system behaviour. If we solve it for one particular set of initial values right from the start, we strive for a *numerical solution*.

Computers yield numerical solutions. This statement is not 100% correct. There are computer programs which yield symbolic solutions. They are called computer algebra systems. While they are very powerful, they cannot find an analytical solution always and obviously do not yield a result if there is no analytical solution. We walk down the numerics route in this course.

Analogous to this distinction of numerical and analytical solutions, we can also distinguish how we manipulate formulae. If you want to determine the derivative $\partial_t y(t)$ of $y(t) = t^2$, you know $\partial_t y(t) = 2t$ from school. You manipulate the formulae symbolically. In a computer, you can also evaluate $\partial_t y(t)$ only for a given input. This often comprises some algorithmic approximations. In this case, you again tackle the expression numerically.

Benefiting from analytical solutions

Whenever you know of an analytical solution of a problem for special cases (configurations), you should first of all check whether your numerical computer code yields the right result for these special cases. It is the simplest sanity test one can think of and the first thing you should do when you code.

Further Reading

- Alain Chenciner: Three body problem. Scholarpedia. 2(10), 2111 (2007)
  http://www.scholarpedia.org/article/Three_body_problem

---

[4] Mathematicians like *analytic functions*. That's something completely different, namely functions that can infinitely often be differentiated in $\mathbb{C}$. We always speak of analytical solutions here. Our "analytical" highlights how the solution has been obtained, i.e. via exact, symbolic manipulation of the equations. The "analytic" in contrast is a property of a function.

> **Key points and lessons learned**

- We have introduced the term ordinary differential equation.
- We know the difference between an analytical and a numerical solution.
- Many problems in science and engineering are given by differential equations. These are models of the real world behaviour. Despite the simplicity of some models, we cannot solve them analytically, i.e. we don't know a solution for these equations.
- The mainstream of scientific computing focuses on techniques how to solve scientific problems numerically.

Wrap-up

The three chapters of Part I give us an idea why we have to study numerical techniques: Many (perhaps most of the) interesting phenomena in science and engineering are phrased as differential equations. Often, no analytical solution to these equations is known. We therefore need numerical codes that give us answers for scenarios we are interested in.

These answers have to drop in fast. Tomorrow's weather forecast is useless if it drops in after 48 hours. Furthermore, many problems become fun and meaningful only if we can make them really large and hence costly. If you can simulate three molecules, then this is interesting. But what people really are interested in are gas volumes hosting millions of molecules. The N-body problem as sketched is relatively simple yet already impossible to solve analytically. We therefore stick to this problem for the majority of this book.

Getting results faster needs us to write code that is aware that performance growth today and in the future is due to an increase in parallelism. We therefore will try to make our N-body code scale. Multiple flavours of parallelism do exist in modern computers.

In the next sections, we will study how floating point calculations are to be phrased such that they are processed effectively on modern machines. We already mentioned that numerical solutions often are approximations. There is more to that: Computers make errors in every single computation, and they inherently cannot represent some numbers. We thus have to study these errors. We have to understand to which degree we can trust the outcome of our calculations.

Part II
How the Machine Works

Floating Point Numbers

4

Abstract

Numbers from $\mathbb{R}$ cannot be handled directly by computers, as they can have infinitely many digits after the decimal point. We have to truncate their representation. Storage schemes with a finite number of digits before and after the decimal point struggle: There are always regimes of values where they work prefectly fine and regimes where they cannot represent values accurately. We therefore introduce the concept of floating point numbers and discuss the IEEE standard. Its concept of normalisation allows us to establish the notion of machine precision, as we realise that the truncation into a finite number of digits equals rounding and introduces a bounded relative error. Since we work with approximations to real numbers, we have to be careful when we compare floating point numbers in our codes.

In this book, we are primarily interested in numbers from $\mathbb{R}$. In the C language, this means `float` and `double`. We use these data types in force calculations, for particle positions, for all kind of equations, but ignore challenges around integer numbers.

Assume our computer represents each number in $\mathbb{R}$ with 10 decimal digits. Let's neglect that there are negative numbers and that computers typically work with a binary system.

We first split up the ten digits into two parts: The first five digits are the digits in front of the decimal point, the remaining five digits are those after the decimal point. As a result, the number 1 is represented as $00001|00000$. I inserted the $|$ to highlight where we decided to have the decimal point by convention. The $|$ is not there in the representation within the machine. Starting from $x = 1$, we can compute $x/2 = (00000|50000)$, $666 \cdot x = 00666|00000$ or $x/100,000 = 0000|00001$. Our number format is however ill-suited to en-

© The Author(s), under exclusive license to Springer Nature Switzerland AG 2021
T. Weinzierl, *Principles of Parallel Scientific Computing*, Undergraduate Topics in Computer Science, https://doi.org/10.1007/978-3-030-76194-3_4

code the numbers $x/100{,}000$ or $x/3$, as we are running out of proper digits after
the decimal point. Very large numbers also pose a problem. Can you quantify
the error we obtain?

In our second attempt as chip designers, we devise a new format where we
sacrifice the first digit. Let the first digit tell us where the decimal point is found.
$4\ldots\ldots\ldots$ implies $4\ldots\ |\ldots\ldots$, i.e. we have the decimal point after
the fourth real digit. With the first digit being 4, the remaining bits encode a
number similar to the scheme before. However, we can now encode numbers
like $x/200{,}000$ as `3000000005`, i.e. `000|000005`. A few questions arise:
(i) What is the smallest number we can write down now? (ii) We obviously still
struggle to represent $1/3$. The best we can do is to write it down as `0333333333`
meaning 0.333333333. What is the error that we get? (iii) How much more
accurate is this than the same number written down in the first format?

Computers use a fixed number of bytes per number; typically 4 or 8; to store values
from $\mathbb{R}$. In theory, machines could work with very flexible formats, where the number
of bits that the machine invests per number is not fixed a priori. However, that would
mean that the memory footprint of codes could suddenly explode, some values remain
"unstorable" (take $1/3$ or π) and it would make the hardware very slow. Therefore,
we work with a finite number of bits and *approximate* "the real thing". With a finite
number of bits, we can encode a finite number of values. $\mathbb{R}$ however is unbound and
continuous, i.e. we can zoom in further and further. Infinitely many values in $\mathbb{R}$ thus
cannot be encoded. This property causes pain.

4.1 Fixed Point Formats

The simplest scheme one might come up with for a machine is a fixed point storage
format. In fixed point notation, we write down all numbers with a certain number of
digits before the decimal point and a certain number after the decimal point. With
four digits (decimal) and two leading digits, we can, for example, write the number
three as `0300`. This means 03.00. Fixed point storage is the first variant I have
sketched in our introductory thought experiment.

We immediately see that such a representation is not a fit for scientific computing.
Let $x = 3$ in the representation be divided by three. The result $x/3 = 1.00$ suits
our data structure. However, once we divide by three once more, we start to run
into serious trouble. Indeed $1/3 \approx 0.33$—which is the closest value to $1/3$ we can
numerically encode—is already off the real result by $0.00\overline{3}$.

You might be tempted to accept that you have a large error. But as a computational
scientist, you neither can accept that most of your bits soon start to hold zeroes, i.e. no
information at all, nor that your storage format is only suited to hold numbers from
a very limited range (basically from 0.01 to 30 and even here with quite some error).

Definition 4.1 (*Relative and absolute error*) Computers always yield wrong results. To quantify this effect, we distinguish the absolute from the relative error. Let x_M be the machine's representation of x. The *absolute round-off error* then is given by

$$e = |x_M - x|.$$

The *relative round-off error* is given by

$$\epsilon = \frac{e}{|x|} = |\frac{x_M - x}{x}|.$$

The reason behind inaccurate representations of numbers is that we work numerically. Each number in the computer is mapped onto a sequence of bits. This sequence is finite. So at one point, we have to cut the bits of the real number off. We introduce an error.

As this is error is rather huge for fixed point notations, and as we have a rather limited range that we can represent, fixed point notations are not used by modern computer hardware anymore. We mainly use it here to introduce our error terminology. However, there has been a time when off-the-shelf computers had not been equipped with hardware compute units for numbers. In that era, game developers, for example, wrote their own fixed point algorithms, as a fixed point number can be realised with two integer values. These number formats suffered from the accuracy and range arguments we made, but for many games they did the job (and we will return to the range argument later, as no number format is immune against all issues). We still encounter fixed point formats occasionally; though mostly outside of the core calculation business. Some scientists are faced with enormous data sets. It is reasonable to think about situations where you can project your numbers onto fixed point numbers, and then use some classic compression algorithm for integers to bring down the memory footprint. Such a process is a *quantization* where ranges of numbers are mapped onto one representative number. This might be inappropriate for the actual compute data, but a good option for visualisation, e.g. The most prominent examples for fixed point techniques I know of are Bresenham's line algorithm—which used to be the number one algorithm to draw a line on the screen–and the rationale behind the JPEG image format.

> Decimal versus binary notation

On the next pages, we use two types of number presentations: decimal and binary. Decimal is what we use in everyday life. We write down numbers like $a_1 a_2 a_3 a_4 . a_5 a_6 a_7 \cdot 10^{a_8 a_9}$, where $a_i \in \{0, 1, \ldots, 9\}$. Technically, this means that we write down numbers as $(a_4 \cdot 10^0 + a_3 \cdot 10^1 + a_2 \cdot 10^2 + a_1 \cdot 10^3 + a_5 \cdot 10^{-1} + a_6 \cdot 10^{-2} + a_6 \cdot 10^{-3}) \cdot 10^{a_9 \cdot 10^0 + a_8 \cdot 10^1}$. Note how we work our way through the digits starting from the decimal point both towards the left and the right, but also for the exponent (a_8 and a_9). If I want to make it very clear that we are using this decimal system, I attach $|_{10}$ to the numbers. If nothing is written down, you may assume that any number is given for a base of 10.

On a computer, a different data representation is used: a binary system. Here, the digit sequence $a_1 a_2 a_3 a_4 . a_5 a_6 a_7$ has a different meaning. First, we use a base of 2 for the scaling of the entries. Second, the a_is are from $\{0, 1\}$ only. Third, we would use $2^{a_8 a_9}$ if we had a scientific notation. In a binary system, digit sequences denote the number $(a_4 \cdot 2^0 + a_3 \cdot 2^1 + a_2 \cdot 2^2 + a_1 \cdot 2^3 + a_5 \cdot 2^{-1} + a_6 \cdot 2^{-2} + a_7 \cdot 2^{-3})$. If required, I will write $a_1 a_2 a_3 a_4 . a_5 a_6 a_7 \cdot 2^{a_8 a_9}|_2$ to make it clear that I'm using a binary base.

We conclude that we search for a scheme where the number of positions after the decimal point is not fixed. This way, we can cover a way bigger value range than with fixed point formats. We are more flexible. At the same time, the new format should run as efficient as possible on our hardware and its memory footprint should be under control. It should work with a fixed, predetermined number of bytes. In computer architecture, there used to be a zoo of such formats. In 1985 however, vendors agreed on the IEEE Standard 754:

4.2   Floating Point Numbers

The IEEE standard makes computers work with *floating point numbers*. The point will be floating around within the number representation. It is conceptionally close to our second variant from the introductory thought experiment: However, we write down the format with an explicit exponent, and we let the exponent determine where the decimal point is. The point does not float around in the bit representation, but it floats around in the real representation. Lets derive this format step by step.

4.2.1   Normalisation

We start our introduction into machine data representations with a simple observation: we need normalisation. It is obvious that $a.bcd \cdot 10^4 = ab.cd \cdot 10^3$. Whenever we write down a number, we have some degree of freedom. Lets exploit this degree of freedom—well, lets actually remove it—and always move the point to the left such that the leading digit is 1. This makes the notation unique, as we work in a binary world. So let $ab.cd \cdot 2^4|_2$ be a number that is not normalised. The same number $a.bcd \cdot 2^5|_2 = 1.bcd \cdot 2^5|_2$ is normalised.

Definition 4.2 (*Normalised number representation*) If a number is written down as

$$\hat{x} = (-1)^{\hat{s}} \left(1 + \sum_{i=1}^{S-1} \hat{x}_i 2^{-i}\right) \cdot 2^{\cdots},$$

it is in its *normalised representation*.

Normalisation is nice to realise (in hardware or software): It requires bit shifts which is something computers are really good at:

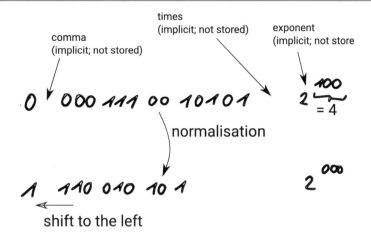

In the upper representations, there are four zeroes to the left of the leftmost digit 1. If we assume that the one and only bit left of the decimal point should be a one, then we have to move all the digits four positions to the left. This is called a bit shift. Before the shift, we had an exponent of $4|_{10} = 100|_2$ which we have to reduce in return.

I emphasise that the discussion of the exponent above is "wrong". We'll discuss that in a minute. However, the principle idea holds: Hardware shifts around the bits, a process we call *normalisation*.

> **Zero**

Zero is the only number we cannot write down in a normalised way, since there is no leading non-zero.

4.2.2 The IEEE Format

Once we write down any (non-integer) number as

$$\hat{x} = (-1)^{\hat{s}} \left(1 + \sum_{i=1}^{S-1} \hat{x}_i 2^{-i}\right) \cdot 2^E \qquad (4.1)$$

the bit sequence is a unique representation of the number x. It consists of S *significant* bits of which one bit is sacrificed for the sign bit $\hat{s}$. The $\hat{x}_i$ are the bits after the decimal point. The significand[1] is supplemented by E *exponent* bits for the exponent. Other names for significand are fraction or mantissa.

Problems result from the fact that we have to squeeze this representation into a finite number of bits. Lets assume that we have 24 bits for the significand plus sign.

---

[1] Storing the significant bits—note the d versus t.

One bit is "lost" for the sign. This leaves us with 23 bits. We get one back, as we do not have to store the leading 1 bit of the normalised representation. We know it has to be there as we have defined normalisation that way. There is no point in storing it. As a consequence, we know exactly how many bits we have available to store the significand. For a 32-bit floating point number, we use 23 bits for all the significant bits right from the decimal point. This is C's `float`. For a 64-bit number, a C `double`, we use 52 bits (Table 4.1).

Definition 4.3 (*Truncation as rounding*) If we take a normalised floating point number of any bit count and squeeze it into the (IEEE) floating point number, we might throw bits away. We truncate the representation such that it fits into our predefined number of bits. Effectively, this is *rounding* or *chopping off*. The other way round, we might have to add bits as we move the significand to the left. In this case, we add either 1s or 0s. Both of them might be the wrong thing to add (compared to the exact math), so we effectively add garbage.

> **Round-off versus truncation terminology**

We find both the term round-off error and truncation error in literature. Round-off emphasises that we might interpret any truncation of the significand or any fill-in of bits as rounding. Truncation emphasises that we, well, truncate the significand if it exceeds our standard. Chopping off a sequence of computations or digits is something we do frequently in scientific computing to approximate an object that is infinite—such as our sequence of digits. Consequently, the verb "truncate" arises in different context. To make things clear, I speak of *rounding* and the round-off error or rounding error whenever we encounter a floating point number.

We obtain a number x_M (note the subscript M for machine precision) which differs from the real number x. x_M is given in machine floating-point precision whereas a real number $x \in \mathbb{R}$ in mathematics has an infinite precision.

In hardware, the table below represents exactly how data is stored: The sign bit comes first, then the exponent. All remaining bits hold the significand. The advantage of such a standardised format is clear: Once we commit to it, we can build hardware to realise it, i.e. vendors can "hardwire" the floating point format and thus make computations with it fast.

Table 4.1 The bit semantics in the IEEE standard

Type	Sign	Exponent	Significand	Total bits
Single	1	8	23	32
Double	1	11	52	64

Definition 4.4 (*Biased exponent*) The *exponent* in the IEEE standard is stored in a *biased form*. That is, we do not store the E from (4.1) directly within our 8 or 11 bits, respectively. Instead, we store $E + 127$ (single precision) or $E + 1023$ (double precision).

Such a shift implies that

1. we can always recompute the real exponent quickly (just substract 127 or 1023, respectively);
2. the value we store is always positive, i.e. we don't have to encode a sign explicitly;
3. we can hold values in the range $-126 \leq E \leq 127$ (single precision) or $-1022 \leq E \leq 1023$ (double), respectively.

With all of these definitions available, we can study a few particular bit sequences within the IEEE standard. For this, we split the range of numbers into certain regions. We start with numbers of the type $\pm 1.00001 \cdot 2^{11111111}|_2$. This is the largest exponent one could think of and a significand that is slightly bigger than one. Though the codes would represent a meaningful number according to our rules, the IEEE standard reserves them:

> **NaN**

If the exponent bits are all set and any bit of the significand is set, too, then the floating point number represents NaN (not a number). As we have a dedicated sign bit, there's a +NaN and a −NaN.

The biggest or smallest, respectively, number that remains now is $\pm 1.0 \cdot 2^{11111111}|_2$. We assign them special semantics, too:

> ∞

If the significand equals one and all the bits of the exponent are set, then the number represents $\pm \infty$ (depending on the sign bit). As we do not store the leading bit in the significand, the code for these two numbers is all zeros in the significand and all ones in the exponent.

Any bit sequence $-1.0 \cdot 2^{11111111}|_2 < x_M < 1.0 \cdot 2^{11111111}|_2$ is a valid number. Starting from the biggest one we we can decrease them further and further. Eventually, we approach $1.0 \cdot 2^{0}|_2$. Note the biased exponent. This is the positive number closest to zero that we can encode. The bitset still would allow us to encode $0.1 \cdot 2^{0}|_2$. Yet, we cannot normalise it anymore, as we run out of digits with the exponent.

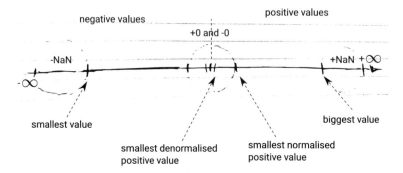

Fig. 4.1 Schematic classification of the different IEEE codes/number types on the number line

> **Denormalised values**

If the biased exponent equals zero and the significand's bits are not zero, we have a denormalised floating point value.

Finally, we have a zero value where the biased exponent is all zeros and the bits in the significand are zero, too. We reserve these codes for the zero value by convention:

> **Signed zero**

The IEEE standard works with signed zeros, i.e. there is $+0$ and -0.

We summarise that there are five types of IEEE numbers (Fig. 4.1): We have not-a-number (NaN) types, $\pm\infty$, normalised numbers, denormalised numbers and two zeroes.

4.2.3   Realising Floating-Point Support

The fixed number of bits per number allows vendors to build bespoke hardware for floating point values. In the early days, floating-point units (FPUs) were shipped as accelerators, i.e. have not been part of the main chip. They have been floating-point co-processors. We had to wait till the 1980s for vendors to integrate the FPU into the CPU.

On the one hand, this is due to the fact that complicated math routines and calculations used to be a specialist domain before that. Today, every smart phone uses image processing, text recognition, and so forth. On the other hand, it follows Moore's logic from Chap. 2. FPUs are very complex and their transistor count today exceeds the number of transistors of other parts of the chip. As a consequence, it used to be economically unfeasible to ship FPUs with every chip. However, having an FPU separate from the CPU means that we have to accept some overhead for them to communicate, i.e. to instruct the FPU and to bring data into the CPU. As soon as a higher transistor count came into reach, it thus was clever to integrate CPU and FPU. There is still some legacy of all of this in modern chips: FPUs often run at a lower frequency than the main chip and have separate access to some memory parts (caches), while high-end chips today feature more than one FPU per CPU.

The IEEE standard is an at-least standard. If you store a binary file with IEEE values, then you know exactly how many bits are invested into every single number. However, there is no guarantee that the data within the chip is held in that precision. Indeed, most FPUs work with a higher precision internally. Intel, for example, spends 80 bits on a double precision value. When we add two numbers, we temporarily have to denormalise. Without the additional bits, we would immediately loose information. It therefore is important that computers internally hold way more bits. When they write values into a file or main memory, then they squeeze them into the IEEE format. As a result, you can get two different simulation results when you once run a simulation without any file dumps and then rerun the simulation but dump its state into a file and reload it.

> ### Return of the floating point formats
>
> IEEE floating point precision is actually not that old. Way into the mid-90s, different vendors implemented their own format. This is not a problem as long as you stick to a single vendor. Yet, once users start to exchange files between different computers, all kind of compatibility issues arise. You can not expect that you can take one number and just push it to another machine. Even worse, if you use the same code, it will yield different answers on different machines.
>
> Around 1985, mankind got the ANSI/IEEE standard. Quickly afterwards, almost all computer architectures implemented this one. The standard defines many things, but the dominant things are single (32 bit) and double (64 bit) precision. These two formats shape our business.
>
> Around 2018, when the machine learning (ML) and neural network hype dominated everything, people started to discuss other precisions. In ML, everything is uncertain and an exact error analysis in most cases is not available anyway, but compute throughput determines how successful a model is in practice. Hardware vendors—GPU manufacturers first, then mainstream processor architects—thus started to introduce half precision, while some mathematicians make the case for beyond-double precision. The "happy" times of only two data formats are over.

4.3 Machine Precision

Floating point numbers in action

Let there be two numbers which we want to add. For this exercise, we assume that we have a floating point format with five significant (post-point) bits, we ignore the sign bit, and we use an unbiased, positive exponent.

$$1.00101 \cdot 2^{10}|_2 + 1.00001 \cdot 2^{00}|_2$$

As the two numbers have a different exponent (they are of different magnitude), the computer's floating point unit first has to bring them to the same exponent format.

$$1.00101 \cdot 2^{10}|_2 + 0.0100001 \cdot 2^{10}|_2$$

Here, it helps that the computer internally works with more significant bits. If *denormalisation* for the right summand exceeds the number of internally available bits, we already lose information. With the two exponents having the same value, we can sum up the significands:

$$1.00101 \cdot 2^{10}|_2 + 0.0100001 \cdot 2^{10}|_2 = 1.0110101 \cdot 2^{10}|_2$$

If this result is now immediately used again by the computer (as part of a larger formula, e.g.) and resides within the CPU's internal memory (register), then we are done. Otherwise, we have to round it, as we have only five significant bits in our toy data format:

$$1.00101 \cdot 2^{10}|_2 + 0.0100001 \cdot 2^{10}|_2 = 1.0110101 \cdot 2^{10} \approx 1.01101 \cdot 2^{10}|_2$$

We introduce an error of $0.0000001 \cdot 2^{10}|_2$.

After any operation—besides some special cases like the multiplication with two—the system will apply some kind of normalisation, i.e. round, and thus "spoil" our result. We therefore know that we are computing with slightly incorrect values from the very moment we run the first arithmetic operation on data.

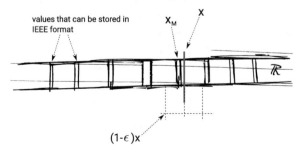

We pretend to work with values x, but the floating point numbers do, for most choices of x or results from calculations, not allow us to represent them precisely.

The format rounds the data to a number that it can represent, and we end up with a machine number

$$x_M \in [x - e, x + e].$$

The number x_M we are actually working with is from an ϵ-environment around the real value x. In this manuscript, we typically write

$$x_M = x(1 + \epsilon)$$

and allow the ϵ to have either sign. Note that we use two different symbols (e and ϵ) as they are different things. They fit to our definition of relative versus absolute error. The sketch supports the statement that ϵ shall have the "worst-case sign". The worst-case magnitude of ϵ is something that depends on the floating point format, i.e. how many significant bits we have. It obviously is bigger for single precision than for double precision.

The latter definition, i.e. the one with the relative error, is useful: If our x is to hold a value around 1,000 but is actually off by 0.1, then this might be acceptable. If our x should store a value around 1 but is off by 0.1, then this is a massive deviation.

Definition 4.5 (*Machine precision*) The *machine precision* for a given floating point format is the upper bound on the relative error.

The above definition is slightly biased: as we assume we have a value which is in the range of what we can represent with our normalised floating point format, i.e. we can "cover" it through our exponent. Another definition of this term goes as follows: The machine precision is the difference between 1.0 and the next number closest to 1.0 that we can represent in our floating point format. Both definitions are the same in the regions we are interested in. The latter eliminates the division from the relative round-off error and we can calculate the machine precision on our computer straightforwardly. It is only a brief source code snippet:

```
double epsilon = 1.0;
double previousGuessOfEpsilon = 1.0;

while ( (1.0+epsilon) != 1.0 ) {
  previousGuessOfEpsilon = epsilon;
  epsilon = epsilon / 2.0;
}
```

We start with a ridiculously high estimate of ϵ and successively reduce it until the code can not spot any difference anymore between 1.0 and $1.0 + \epsilon$. When we end up in this situation (while loop terminates), we unfortunately just have lost the last meaningful bit of ϵ already. Therefore, we always backup the previous value of ϵ which is also the value we should plot in the end.

Machine precision in C

We have a machine precision of around 10^{-7} for C's `float` (single precision) and 10^{-16} for `double` (double precision).

4.4 Programming Guidelines

Being aware of floating point inaccuracies, we may assume

$$x(1 - \epsilon) \le x_M \le x(1 + \epsilon).$$

This has implications for our coding:

Comparisons

Never compare two floating point numbers with the equals operator ==. In C, this operator compares the left and right side bit-wisely. As we work with imprecise numbers subject to round-offs, we always have to write code which checks against "reasonably close".

Whenever we compare two numbers, we can basically assume that they are the same if their (relative) difference is small. I always use statements like

```
if (
  std::abs(x-y)/std::max(std::abs(x),std::abs(y)) < 1e-10
) {
    // x and y are equal in floating point precision
}
```

or I use a comparison of the absolute values if I have an idea about the value quantities:

```
if ( std::abs( x-y ) <= 1e-10 * magicScaling ) {
    // x and y are equal in floating point precision
}
```

If you want to make it 100% watertight, you have to modify the range comparisons, too: Let $x < y$, $x <= y$, $x == y$ be three expressions that you evaluate up to a precision of $\tilde{z}$—so the $\tilde{z}$ is either an absolute or a relative delta. As you use $x ==_M y \mapsto |x - y| \le \tilde{z}$ in your code in line with the rules above, a consistent logic realises the range comparisons as $x <_M y \mapsto x < y - \tilde{z}$ and $x \leq_M y \mapsto x \leq y - \tilde{z}$. Otherwise, a value x could both be smaller and equal to a value y.

Overflows

If you add two floating point numbers and their result exceeds the range that can be represented, you will get ∞.

The process described above is called *floating-point overflow*. On the one hand, this behaviour is re-assuring: Different to integer numbers, nothing weird happens such as a sudden sign flip. On the other hand, you have to be very careful (and check yourself somehow) if you work close to the maximum value that can be encoded by your number format.

NaNs

Your computer will not check for NaNs.

Feel free to divide by zero, e.g., but be prepared to bear the consequences. You can always check if a datum[2] is valid by adding

```
if (x==x) {
   // x==x will fail for NaNs
}
```

Two NaNs are by definition never the same. As so often, C++ has a nicer way to write down things:

```
if (std::isnan(x)) {
   // x==x will fail for NaNs
}
```

---

[2] From Latin, I consider *datum* to be singular and *data* plural.

and the language also offers `std::isfinite`, `std::isinf` and `std::isnormal`, i.e. you can even check whether a floating point number can be represented in a normalised way.

Denormalised numbers

Avoid working with denormalised numbers.

Modern machines can work with denormalised numbers. There are three problems however: First, working within the denormalised regime is slow. Modern chips need significantly more time per floating point operation if you feed them denormalised numbers.

Second, a floating point number has 23 or even 52 bits to discriminate different values from each other. If your first 20 bits, for example, are zero as you work with denormalised data, then only 3 or 32 bits, respectively, hold actual information. That is, the information density of your values is very limited. The effect when you lose the very last meaningful bit and drop from a denormalised number to the real zero is called *floating point underflow*. Underflows can occur from two directions, i.e. into $+0$ and -0.

Finally, some IO/data formats (such as VTK) usually work with reduced precision and in particular struggle to handle denormalised values. You will get visual garbage if you use denormalised data, or your data files might become unreadable. Therefore, I often round to values significantly away from machine precision before I actually write them into a file. Note that C/C++'s terminal output typically also truncates, i.e. you won't get full floating point accuracy here either.

Further reading

- Jack E. Bresenham: *Algorithm for computer control of a digital plotter*. IBM Systems Journal. 4(1), pp. 25–30 (1965) https://doi.org/10.1147/sj.41.0025
- Gregory K. Wallace: *The JPEG still picture compression standard*. IEEE Transactions on Consumer Electronics. 38(1), pp. xviii–xxxiv (1992)
- Markus Wittmann, Thomas Zeiser, Georg Hager, Gerhard Wellein: *Short Note on Costs of Floating Point Operations on current x86-64 Architectures: Denormals, Overflow, Underflow, and Division by Zero*. arXiv:1506.03997 (2015) https://arxiv.org/abs/1506.03997
- David Goldberg: *What every computer scientist should know about floating-point arithmetic*. ACM Computing SurveysMarch. 23(1), pp. 5–48 (1991) https://doi.org/10.1145/103162.103163
- Stanley P. Lipshitz, Robert A. Wannamaker and John Vanderkooy: *Quantization and Dither: A Theoretical Survey*. Journal of the Audio Engineering Society. 40(5), pp. 355–375 (1992) http://www.aes.org/e-lib/browse.cfm?elib=7047

> **Key points and lessons learned**

- We have defined the terms relative versus absolute error. We will reencounter these terms frequently throughout this book.
- We know the storage format of IEEE floating point numbers.
- We can perform simple calculations with floating point numbers. This implies that we can manually normalise and denormalise.
- We have introduced the term round-off error, and we can compute it for a given piece of data (cmp. normalisation and IEEE format's bit length).
- We are familiar with the ideas behind NaN and $\pm\infty$.
- We have discussed a couple of rules of thumb for what to take into account when we work with floating point numbers.

A Simplistic Machine Model

5

Abstract

The von Neumann architecture equips us with an understanding how a computer processes our program code step by step. Our brief review of the architecture highlights the role of registers for our numerical computations. We will augment the straightforward definition of a register later to realise vector processing, i.e. the first flavour of hardware parallelism, while the architecture review provides clues why many numerical codes struggle with performance flaws and bottlenecks on our CPUs.

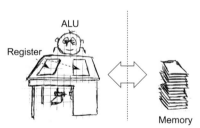

Assume we have to compute the scalar product $f(x, y) = x^T y$, i.e.

$$f = x_1 \cdot y_1 + x_2 \cdot y_2 + x_3 \cdot y_3 + x_4 \cdot y_4 + \cdots$$

on a couple of sheets of paper subject to the following constraints:

- We cannot memorise any data. All results and intermediate results have to be written on sheets of paper.
- Every single sheet of paper can hold only one number at a time.
- We work at our desk. Only two sheets of paper fit onto this desk.

© The Author(s), under exclusive license to Springer Nature Switzerland AG 2021   49
T. Weinzierl, *Principles of Parallel Scientific Computing*, Undergraduate Topics
in Computer Science, https://doi.org/10.1007/978-3-030-76194-3_5

- In the room in front of our office with the desk, we can store an arbitrary number of sheets. They are enumerated.

How can we compute f in such an environment?

Today's computers in principle all evolve from one common blueprint. It is called *von Neumann architecture*. There are a few other fundamental machine architectures that have been published earlier and differ by nuances (there is a Harvard architecture and a Zuse architecture, for example), but we stick to von Neumann here.

In the von Neumann architecture, a (minimalist) computer consists of few building blocks. Each part of the machine, i.e. each block is responsible for a particular job.

Definition 5.1 (*von Neumann architecture*) The *von Neumann architecture* consists of four parts which are relevant for us: A control unit is the mastermind or conductor which orchestrates all other machine parts. An arithmetic logic unit (ALU) is the only part of the machine which can run calculations. A main memory holds data. A bus connects the ALU and the control unit with the main memory. It also connects the machinery to other devices such as the keyboard or the screen.

The control unit and the ALU are often integrated into one piece. The whole machine works in steps. It is pulsed. Per step, its components perform at most one operation. If you specify your machine with GHz ($1\,\mathrm{Hz} = 1/\mathrm{s}$), this tells you how many of these basic steps are done by the computer per second, though modern machines use different frequencies for the different machine parts. The arithmetic unit is the working horse of the computer. Yet, it is surprisingly primitive: it can, for example, add two numbers or subtract them. It can also compare two numbers. This is the reason for the term arithmetic logic unit (ALU). In the memory, all data of a program are stored as a long sequence. The enumeration starts from 0.

The ALU has registers. A register is its own (fast) local memory which can hold one datum. There is a limited number of registers which is small compared to the main memory, as registers are expensive to manufacture and power. As a consequence, we have to bring in data into the ALU before we use them, and then we have to move it back into the memory to free registers for the next calculation.

5.1 Blueprint of a Computer's Execution Workflow

The description so far is a sketch of the arrangement of technological components within a machine (Fig. 5.1). It is a static description. We next run through one example how a code is, schematically, executed on such a machine. It is a simple code snippet calculating a scalar product. The whole thing is a thought experiment, i.e. a simplified, artificial von Neumann machine serves as testbed. Our code looks as follows:

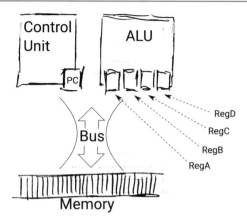

Fig. 5.1 Schematic sketch of a von Neumann architecture: We have three components (ALU, Control Unit and memory) which are connected via a bus. The main memory holds both the machine code and the application's data. Its locations are enumerated. In each step, the control unit reads from the position held in PC from the main memory. The code (instruction) there tells it what do do next. It then loads values into the ALU, or tells the ALU to compute something, or moves data back from the ALU registers into the main memory

```
double f = 0.0;
double x[...]; // our input array
double y[...]; // the other input array
for (int i=0; i<I; i++) {
  f += x[i] * y[i];
}
```

Definition 5.2 (*Machine code*) The above code is *high level code*, i.e. it is nothing a computer can directly execute. A computer needs much simpler instructions which the control unit can directly use to instruct the von Neumann machine parts. Such a *low level code* is also called a *machine code*.

Machine code is something that is produced by your C compiler. Whenever you type in gcc for example, your compiler takes the high-level C code and produces machine code for your computer. This machine code can subsequently be executed on your machine. As the name suggests, it is specific for your machine (type). If you copy it to a different architecture, it might not work. However, many compilers allow you to tell them that you want to produce machine instructions for another machine rather than the one you are compiling on. This is called *cross-compilation*. Writing code for GPGPUs can be read along these lines: You write down the code on your machine, and you then ask the compiler to produce machine code for the GPU. You usually do not translate on the GPU itself.

To understand what machine code looks like, we next clarify what kind of ALU we have in our made-up machine. We have to know what it is capable to do and how it works. Though there are significant architectural differences, the core principles of ALU/von Neumann architectures as we use them in the artificial machine are similar to those in real architectures. For our code snippet, the ALU has to be able to do three things: add two numbers, multiply two numbers and compare a number to another number. Let the ALU have only five storage places. Recall the office desk on which we can fit few sheets of paper only: Each sheet can hold one number. The ALU can, if instructed to do so by the controller, compare two sheets and write its result to either of them again, or it can combine two sheets' values by a multiplication or addition and write the result back onto one of the sheets. We call these "sheets" *registers*.

It is the responsibility of the control unit to ask the bus to bring the actual x, y and f values into the ALU registers. After that, the control unit has to tell the ALU to add, compare or multiply the two values. The result is deposited by the ALU in one register. Finally, the controller has to instruct the bus to bring the result back from the registers into the memory; to free the former for further work. With this machine (code) idea, a part of the machine operation sequence for our high level code looks as follows:

```
load mem... into RegA
load mem... into RegB
add RegA and RegB into RegA
store RegA into mem...
```

We load the data from a particular *address* in memory into one of our registers (here called RegA). The address is a positive number identifying a memory location. The location holds the value of a. We then do the same thing for the other variable and RegB. After we have added the two numbers, the result ends up in RegA. We finally write this result back to main memory, and RegA becomes available for the next calculation.

A simple C instruction like $f = x[\ldots] + y[\ldots]$ translates into at least four machine code steps, where we have to be careful which data is held in which register at which point in time. Identifying a suitable set of registers and getting the memory moves right is a tricky challenge mastered by your compiler. The underlying process is called *register assignment*.

Our machine code for our simple von Neumann architecture consists only of a few instruction types. It itself is a sequence of codes stored in the main memory just as the data itself.[1] Each step within this code is either

- move a constant into a register. In our simple case, let's have four registers: RegA, RegB, RegC, RegD;
- move one piece of data from a position in main memory to a register. Where to take the data from, i.e. its *address*, is held in one of the registers, too;
- move data stored in RegA, RegB, RegC or RegD to the memory location specified by the content of RegA, RegB, RegC or RegD;[2]
- add or multiply the content of any of the two registers. The result becomes available in one register as specified;
- compare any register to any other register. If the two are equal, write a non-zero number into one of the registers. Otherwise, store zero;
- if a register holds zero, skip the next line of the instruction.

The control unit has a special register itself, the *program counter* (PC). After a machine code step has terminated, this counter is incremented; unless the code explicitly overwrites the PC register's value or encouters the skip command which increases the PC by two. At the startup, the PC register points to the very first instruction of your code. By manipulating the PC, a code can jump around in the instruction stream, as the controller always reads the next instruction from where the PC points to.

Machine code has no variables or even arrays. It works only with memory addresses. That is, the notion of an array a and its elements a[i] is a high level language concept. In reality, an array materialises as a long consecutive sequence of numbers in main memory. For our simple machine, we can assume that constants, i.e. any hard-coded number, can be stored within the instruction stream. We can write things like move 14 into RegA. For real machines, this is often possible for integers only. For floating point numbers, it is the responsibility of the compiler to create a big table of floating point values embedded into the machine code. Instead of moving a magic constant into a register, we can tell the bus to move a particular memory location from the lookup table into a register.

---

[1] The fact that we do not distinguish code and instructions within the memory—they are technically both just number sequences and the software has to interpret them either as data or instruction codes—is a built-in security problem. Malware can alter code by pretending it wrote data into the memory.

[2] This is where the machine truncates the bit sequence of the floating point number into the IEEE format (cmp. Sect. 4.2.3): If the ALU uses 80 bits per double internally for example, the number is now reduced to 64.

5.2 A Working SISD Example

Lets assume that the initial memory configuration for our example with I=10 looks
like this:

```
00: load 50        18: y[8]                      RegD into RegC
    into PC        19: y[9]           58: load mem[20] into RegD
01: x[0]           20: f              59: add RegC and RegD
02: x[1]           21: i                  into RegD
03: x[2]                              60: store RegD into mem[20]
04: x[3]           [...]              61: load 1  into RegC
05: x[4]                              62: load mem[21] into RegD
06: x[5]           50: load 0  into RegD   63: add RegA and RegC
07: x[6]           51: store RegD into     into RegA
08: x[7]               mem[20]         64: add RegB and RegC
09: x[8]           52: store RegD into     into RegB
10: x[9]               mem[21]         65: add RegD and RegC
11: y[0]           53: load 1  into RegA    into RegD
12: y[1]           54: load 11 into RegB  66: store RegD into mem[21]
13: y[2]           55: load mem[RegA]  67: load 10 into RegC
14: y[3]               into RegC       68: compare RegD and RegC
15: y[4]           56: load mem[RegB]     into Reg D
16: y[5]               into RegD       69: if equal skip next line
17: y[6]           57: mult RegC and   70: load 54 into PC
```

This is a very primitive computer architecture. Yet, with only few different instructions, it can compute a scalar product:

- We assume that the memory regions 01–10 hold entries of the array x, while 11–19
  holds y.
- 20 and 21 are spare memory locations that will hold variables later on. I wrote in
  i and f, but initially these memory locations hold whatever has been there before:
  garbage from a code's point of view.
- The first thing we do is that we make the controller jump to location 50 in memory.
  This is where the actual machine code (algorithm) starts. 00 is the code entry point,
  01–49 is the data region and everything ≥50 holds machine code.
- We load a zero into RegD and use this value to initialise both the storage locations
  of i (the loop counter) and the result f.
- We initialise RegA and RegB. These two registers will always point to the current
  entry of the array x or y, respectively.
- We load the current array elements into the registers RegC and RegD, multiply
  them (result is kept in RegC), fetch the content of f, add the multiplication's
  result, and eventually write back the new value of f into the memory location 20.
  This sequence is the loop body.
- We load the value 1 into RegD and use it to increment the two pointers as well
  as the variable i. This is the pointer increment within the high level code's for
  statement.
- We compare i to 10. If it is equal, we skip the subsequent line. This is the end of
  the algorithm. If it is not equal (yet), we jump back to 54 to run through the code,
  i.e. the loop body of our C code, once again.

In subsequent chapters, we will sophisticate the architecture further and, hence, make the machine faster. At the same time, we will work exclusively with C, i.e. a high level programming language. To understand what is going on "under the hood", it is however important to realise how simple high level constructs (such as a `for` loop) are translated into code sequences with register comparisons and register manipulations.

Definition 5.3 (*Single Instruction Single Data (SISD)*) The plain machine as introduced here is a *single instruction single data* machine (SISD). It runs one single instruction at a time. It furthermore runs this instruction on a single piece of data or a pair of entries feeding into a binary operator, respectively.

The term SISD has been coined by the scientists Michael J. Flynn and is part of the so-called *Flynn's taxonomy*. We return to this taxonomy multiple times.

5.3   Flaws

There are obvious flaws in our computer design; and indeed today's machines. Two are important at this point.

For computations such as our vector addition, we have to pipe a stream of entries through the registers. Our ALU has to be served by the memory before it can trigger its actual computation.

Definition 5.4 (*Memory-bound*) Nowadays ALUs are fast, while the memory can not keep pace. We end up in situations where the ALU has to wait for the memory most of the time. Such a code is *memory-bound*.

Its runtime is not determined by the sheer amount of computations but by the capability of the memory subsystem. Many important codes today suffer from memory-boundness.

Definition 5.5 (*Compute-bound*) A code where the performance is dominated by the number of computations that an ALU can do is *compute-bound*.

Compute-bound codes are rare. Memory-boundness is almost intrinsic to the von Neumann architecture: We clearly separate the data storage (memory) from the compute facilities. This clear separation of concerns—a functional decomposition—comes at a price: data movements. Whenever we issue an instruction, we have to wait for the data to arrive. This latency (waiting for data to arrive before we can do anything) is often labelled as the *von Neumann bottleneck*[3]. Bandwidth issues are

---

[3] This is the price to pay for fame: the most important technological achievements might be named after you, but then the flaws tied to it inherit your name, too.

another flavour of this bottleneck. Data movements are not only problematic from a performance point of view. It is memory movements that are responsible for a major part of the power budget, not the actual computations.[4]

A second implication of our SISD/von Neumann design is that it is intrinsically sequential. There is no parallelism in there so far. To extend this machine model, we will return to our desk metaphor. But that's next …

Further reading and historic material

- John von Neumann: *First Draft of a Report on the EDVAC.* Technical report (1945)
- Michael J. Flynn: *Some Computer Organizations and Their Effectiveness.* IEEE Transactions on Computers. C-21(9), pp. 948–960 (1972) https://doi.org/10.1109/TC.1972.5009071
- Randall Hyde: *Write Great Code—Understanding the Machine: Volume 1.* No Starch Press (2004)

> **Key points and lessons learned**

- The four core components of the von Neumann architecture are control unit, ALU, bus and memory.
- A register is the ALU's internal memory (scratchpad) for computations. There is a very small number of registers.
- SISD means single instruction single data. SISD architectures can compute one operation on one piece of data (or one tuple of data for binary operations) at a time. The term is part of Flynn's taxonomy.
- Before we can compute anything, we have to move the required values from the main memory to the registers. Codes that spend the majority of their time moving data in and out are called memory-bound.
- Traditional efficiency metrics often assume compute-bound codes.

---

[4] With a theoretical computer science background, you will characterise algorithms as $\mathcal{O}(n)$ or $\mathcal{O}(n \log n)$ or …where n counts the algorithm's operations. In reality, there might be an additional $\mathcal{O}(m)$, $\mathcal{O}(m \log m)$, …term paraphrasing the memory movement cost. It is not clear a priori which term dominates the actual runtime.

Wrap-up

There are two fundamental principles of computers that drive our studies: Computers have to work with finite representations of data, and computers work with a small number of simple unary and binary operations in a clocked way. These are the two topics in Part II. As our computer can not work with $\mathbb{R}$, it works with an approximation of numbers from $\mathbb{R}$, rounding will become our steady companion throughout the text. The runtime of a code (manipulating data from $\mathbb{R}$) will be determined by the number of binary and unary steps within the machine code produced by the compiler.

With these two fundamental principles at hand, we can next study both the arising errors of a code and its computational efficiency. We can start to do some number crunching. Our goal is, on the one hand, to obtain a good understanding of the underlying principles needed to dive into more details in further literature. On the other hand, we want to start to write down some "rules of thumb" or recommendations: Where do we need extra care when we write code?

Different to classic programming, numerical codes always give the wrong results even when they are correct. It is the approximation with IEEE precision which introduces errors. We have to understand how big these errors can grow for particular cases. Can we trust the prediction how the particles in our simulation move through the domain? On top of this, some ways to write down basic calculations are horribly slow while others run extremely fast. It is important to understand reasons for this, so we can deliver code that is also of high non-functional quality, i.e. speed. So far, we know that counting the number of computations - such as calculations of particle-particle distances - is flawed, as the data first has to be moved from the memory into the registers if it does not yet reside there. Further to that, the type of operation on sets of registers makes a difference!

Part III

Floating Point Number Crunching

Round-Off Error Propagation

Abstract

Round-off errors creep into our calculations with every non-trivial floating point instruction. We introduce a formalism how to analyse round-off errors in implementations, and discuss when and why numerical calculations suffer from a lack of associativity, cancellation or accumulation stagnation.

We configure our model problem from Chap. 3 as intergalactic pendulum: Let the three bodies align along the x-axis at the position $p_{x,1} = 0$, $p_{x,2} = 1$ and $p_{x,3} = 2$ (that's the x-position of particle 1, the x-position of particle 2, and the x-position of particle 3). Only the middle one is free to move according to our Eqs. (3.1), (3.2) and (3.3). $m_1 = m_2 = m_3$, i.e. all three objects have the same mass. We now run two experiments—preferable in your mind first and then as simulation:

1. The left and right object are stationary while the middle one is subject to an initial velocity $v_2 = (0, 1, 0)^T$.
2. Let object 1 (the left one) have a prescribed velocity $v_1 = (-\tilde{v}, 0, 0)^T$, while object 3 has the velocity $v_3 = (\tilde{v}, 0, 0)^T$. They move apart with a tiny $\tilde{v}$.

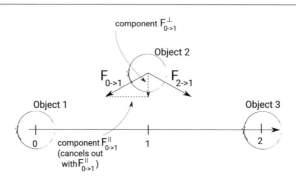

The whole system should then behave like a huge pendulum or spring. As the two outer objects are prescribed and the free object is right in the middle, the forces pull the middle object back whenever it "leaves" the axis the three objects are initially aligned on. The free object oscillates: the left guy pulls it closer, but the right guy does the same. All the "horizontal" pulls cancel out.

The lack of any friction explains why the oscillation carries on and on and on. Indeed, its amplitude remains invariant over time in the first experiment. In the second experiment, the amplitude of the oscillation becomes bigger over time. The left and right object move apart. They are the spring anchors. Therefore, their forces upon the free object become smaller and smaller.

The interesting thing to us is not that the amplitude grows. The surprising observation is that the middle object suddenly goes crazy: It does not remain in the middle, but after a while starts to roam around in the domain.

Our setup starts with perfect oscillations. There are two forces $F_{0 \mapsto 1}$ (force of 0 on 1) and $F_{2 \mapsto 1}$. Both decompose into a force parallel to the axis connecting 0 and 2 and a force orthogonal to this axis:

$$F_{0 \mapsto 1} = F_{0 \mapsto 1}^{\perp} + F_{0 \mapsto 1}^{\parallel} \quad \text{and} \quad F_{2 \mapsto 1} = F_{2 \mapsto 1}^{\perp} + F_{2 \mapsto 1}^{\parallel}.$$

The former two partial forces cancel out, i.e. $F_{0 \mapsto 1}^{\parallel} + F_{2 \mapsto 1}^{\parallel} = 0$, while the two partial forces orthogonal to the axis sum up. They push object 1 back towards the line connecting 0 and 2 whenever it leaves the line due to the initial velocity. In our example, they are even the same: $F_{0 \mapsto 1}^{\perp} = F_{2 \mapsto 1}^{\perp}$.

After a while, the central particle ends up slightly off the centre if 0 and 2 move away from each other. $F_{0 \mapsto 1}^{\parallel}$ and $F_{2 \mapsto 1}^{\parallel}$ are scaled with the inverse of the distance of object 0 and 1 or 0 and 2, respectively. When we compute this distance, we get a slight error due to the scaling's normalisation and rounding. Therefore, $\left(F_{0 \mapsto 1}^{\parallel} \right)_M \neq - \left(F_{2 \mapsto 1}^{\parallel} \right)_M$. We introduce a tiny acceleration parallel to the line between 0 and 2, and consequently move object 1 slightly. Once it has left its central oscillation path, the horizontal forces do not cancel out anymore, so the "wrong" acceleration starts to grow in our simulation. We obtain a weird trajectory rather than a perfect oscillation. Our simulation becomes totally wrong (and neither matches any theory nor any properly done experiment). We now want to ask ourselves where these small deviations came from in the first place. They navigated us into this unsatisfactory situation.

We first formalise our considerations. Let $F_{0\mapsto 1}^{\parallel}$ and $F_{2\mapsto 1}^{\parallel}$ be the horizontal forces in the initial configuration. We know $\left(F_{0\mapsto 1}^{\parallel}\right)_M = -\left(F_{2\mapsto 1}^{\parallel}\right)_M$ bit-wisely as object 1 stays in the middle of 0 and 2 if the latter two objects do not move. As 0 and 2 move away from each other, object 1 experiences forces

$$s \cdot F_{0\mapsto 1}^{\parallel} \quad \text{and} \quad s \cdot F_{2\mapsto 1}^{\parallel} \quad \text{with } s < 1.$$

Due to the machine's finite precision, we cannot represent this value on our machine accurately as s decreases over time:

$$\left(s \cdot F_{0\mapsto 1}^{\parallel}\right)_M \neq s \cdot F_{0\mapsto 1}^{\parallel} \quad \text{and} \quad \left(s \cdot F_{2\mapsto 1}^{\parallel}\right)_M \neq s \cdot F_{2\mapsto 1}^{\parallel}.$$

Even worse, both sides might round the force to the next slightly bigger value. If we had centred our setup around 0, then both forces would be rounded the same way. However, we have shifted the setup, and we know that IEEE does not round two values the same way, i.e. add or remove the same amount to fit the values into IEEE, if they have a different magnitude. With our definition of machine precision, we can assume

$$\left(s \cdot F_{0\mapsto 1}^{\parallel}\right)_M \approx (s \cdot F_{0\mapsto 1}^{\parallel})(1+\epsilon_{0\mapsto 1}) \quad \text{and} \quad \left(s \cdot F_{2\mapsto 1}^{\parallel}\right)_M \approx (s \cdot F_{2\mapsto 1}^{\parallel})(1+\epsilon_{2\mapsto 1}).$$

The $\epsilon_{0\mapsto 1}$ and $\epsilon_{2\mapsto 1}$ are some values whose magnitudes are in the order of the machine precision ϵ. Now we sum up the rounded values and keep in mind that the object 0 is left of the coordinate axis' origin, i.e. this force component is negative.

$$\begin{aligned}(s \cdot F_{0\mapsto 1}^{\parallel})(1+\epsilon_{0\mapsto 1}) + (s \cdot F_{2\mapsto 1}^{\parallel})(1+\epsilon_{2\mapsto 1}) = {}& (s \cdot (F_{0\mapsto 1}^{\parallel} + F_{2\mapsto 1}^{\parallel})) \\ & + s \cdot F_{0\mapsto 1}^{\parallel} \cdot \epsilon_{0\mapsto 1} + s \cdot F_{2\mapsto 1}^{\parallel} \cdot \epsilon_{2\mapsto 1}.\end{aligned}$$

The $F_{0\mapsto 1}^{\parallel} = -F_{2\mapsto 1}^{\parallel}$ (in exact arithmetics) makes the first term drop out, but the other two terms do not cancel as $\epsilon_{0\mapsto 1} \neq \epsilon_{2\mapsto 1}$. It could, for example, happen that the left value is rounded towards the next smaller value, while we have rounded the other value to the next bigger one along the number line. After all, we have to "rasterise" the unknowns into IEEE precision. With $|\epsilon_{0\mapsto 1}| \leq \epsilon$ and $|\epsilon_{2\mapsto 1}| \leq \epsilon$, the right term's size is bounded by around 2ϵ; but it does not disappear.

If we work with floating point numbers in a computer, the permanent denormalisation/normalisation over a finite bit field introduces round-off errors which propagate through our calculations. The results become biased. The fewer bits we have for our values, the more biased we are. This is something we have to accept. We however can not accept that all of these tiny errors start to blow up unheededly. Our goal next thus is to establish a formalism how to study the error propagation.

6.1 Inexact Arithmetics

Since we know that even our most basic arithmetic operations are "wrong", we should write a formula, once it is translated into source code, as

$$f = x + y \Rightarrow f_M = round(x + y).$$

In a large calculation, every single term therein should be replaced with the above rounding wrapper. Wherever we write $x + y$ (or another operation) in our code, we will end up with something slightly off the correct result. To highlight this, I write $x +_M y$. The subscript M denotes that this addition is done in machine precision, i.e. it will, for most cases, not be exact. It is a shortcut for wrapping everything into the $round(\ldots)$ function from above.

We now can use the knowledge from (4.3): We know where the round-off error comes from. The actual calculations suffer from the number (re-)normalisations and the rounding and thus calculate $(x + y)_M$ instead of $(x + y)$. I usually prefer to write $x +_M y = (x + y)_M$ to safe a few brackets—I am not a LISP programmer after all. We end up with

$$(x + y)_M = x +_M y = round(x + y) = (x + y)(1 + \epsilon).$$

The ϵ is a worst-case estimate always picking the sign least favourable for us. So it could be smaller or bigger than zero. It could also be exactly zero, but we cannot count on this. We work, on purpose, quite sloppy here. In a more formal world, we should multiply the real result with $(1 + \tilde{\epsilon})$ with $|\tilde{\epsilon}| \leq \epsilon$ from our machine precision definition. But we keep notation simple. Our goal is not to construct a watertight formalism. We want to understand the implications and issues for us as programmers.

If we have two operations $x +_M y \cdot_M z$ in a row, we'd have to use two epsilon values ϵ_1 and ϵ_2. One for the addition and one for the multiplication. We however ignore all of this and again work with only one generic ϵ (read "$\pm$ the machine precision" as worst-case). This allows us to introduce few simple compute rules:

$$\begin{align}
x +_M y &\Rightarrow (x + y)(1 + \epsilon) \tag{6.1}\\
x -_M y &\Rightarrow (x - y)(1 + \epsilon)\\
x \cdot_M y &\Rightarrow (x \cdot y)(1 + \epsilon)\\
x /_M y &\Rightarrow (x/y)(1 + \epsilon).
\end{align}$$

This formalises what happens in our ALU in a single step. We furthermore simplify our business and neglect all higher order terms:

$$\epsilon \cdot \epsilon \Rightarrow 0,$$

i.e. assume that ϵ is really small, so the product of two ϵ is something we can safely ignore. Finally, we introduce

$$\epsilon + \epsilon \Rightarrow 2\epsilon \quad \text{and}$$
$$\epsilon - \epsilon \Rightarrow 2\epsilon,$$

as we don't know the sign of the errors ϵ. If we add two ϵ, the maximum error doubles. If we subtract one ϵ from the other, we have to assume that they have different signs and thus eventually also add up. Finally,

$$x -_M y = \epsilon$$

if $x \approx y$. This is an effect we call *cancellation*:

Definition 6.1 (*Cancellation*) If we subtract two values x and y from each other and if x and y are almost the same (up to machine precision), normalisation makes us end up with lots of garbage bits. Scientifically correct, we speak of a *loss of significant bits*: The additional bits that we insert when we renormalise the result do not carry any semantics. With x (and y) we might have 64 or 32 bits with a real meaning. If x and y are of the same size $x - y$ will yield a result where very few bits actually mean something, i.e. are significant. This effect is called *cancellation*.

> **Error cancellation**

For many algorithms, the round-off errors of subsequent computations cancel out most of the time. However, if we run an error analysis, we have to be prepared for the worst.

This worst-case analysis is done in the tradition of Wilkinson, and that's from the 1960s. In reality, it is usually way too pessimistic. Some values are slightly too big, some too small. If we add them up, the resulting error does not amplify but more or less cancels out. Modern error analysis models this behaviour by making ϵ a random distribution, e.g.

6.2 Round-Off Error Analysis

With a formal system at hand, we can analyse longer calculations. This analysis describes *how a computer realises a particular source code*. Our activity consists of two steps: We first write down the calculation as done on the machine and replace all arithmetic operations by their machine counterpart. In a second step, we use the rules from (6.1) and thereafter to map the machine calculations onto "real" mathematical operators including error thresholds. The first step is purely mechanical. The second step requires us to think about potential variable content.

I demonstrate this agenda via the summation of three variables. Technically, each compute step would require us to work with an ϵ of its own:

$$
\begin{aligned}
x +_M y +_M z &= (x + y)(1 + \epsilon_1) +_M z \\
&= ((x + y)(1 + \epsilon_1) + z)(1 + \epsilon_2) \\
&= (x + y) + (x + y)\epsilon_1 + (x + y)\epsilon_2 + (x + y)\epsilon_1\epsilon_2 + z + z\epsilon_2
\end{aligned}
$$

For each calculation, we have another error. As discussed before, we use a pessimistic upper bound and hence use a generic ϵ as worst case, i.e. we set $\epsilon_1 = \epsilon_2 = \epsilon$. Furthermore, we eliminate higher order terms and end up with

$$
\begin{aligned}
x +_M y +_M z &= ((x + y)(1 + \epsilon) + z)(1 + \epsilon) \\
&= (x + y + z + x\epsilon + y\epsilon)(1 + \epsilon) && (6.2) \\
&= x + y + z + x\epsilon + y\epsilon + x\epsilon + y\epsilon + z\epsilon + x\epsilon^2 + y\epsilon^2 \\
&= x + y + z + x\epsilon + y\epsilon + x\epsilon + y\epsilon + z\epsilon \\
&= x(1 + 2\epsilon) + y(1 + 2\epsilon) + z(1 + \epsilon). && (6.3)
\end{aligned}
$$

We reiterate that this is not a precise formula in a mathematical sense: Per term, the ϵ accepts the role of a bad guy and switches sign and magnitude just as appropriate. The equals sign also is really a $\approx$ as we drop the higher order terms. All of this has been known before. There are however a few implicit assumptions that we used and should articulate explicitly:

> **Round-off error analysis**

Round-off error analysis is a straightforward set of formal activities:

1. Replace all operations of the machine with machine operations ($+$ becomes $+_M$) but continue to assume that all input data are exact: we do not work with an x_M instead of x as input, but we calculate the outcome $f_M(x)$ of a function rather than $f(x)$.
2. Replace the machine operation with their ϵ-parameterised counterpart. You might need case distinctions for different value ranges!
3. Throw away higher order terms. That is, terms like ϵ^2 simply are dropped. If you run into something alike $\frac{1+\epsilon}{1+\epsilon}$ you assume that that the ϵ in the enumerator is large and the one in the denominator is small. So we can replace $\frac{1+\epsilon}{1+\epsilon} = \frac{1+\epsilon}{1-\epsilon} = \frac{1-\epsilon+2\epsilon}{1-\epsilon} = 1 + \frac{2\epsilon}{1-\epsilon} \leq 1 + 3\epsilon$.

6.3 Programming Recipes

The simple formalism yields some interesting insight into the way we should program:

Associativity

We see directly from (6.3) that

$$(x +_M y) +_M z \neq x +_M (y +_M z).$$

Associativity

With machine arithmetics, associative operations are not associative anymore.

Associativity and worst-case analysis

Both the worst-case character of our analysis and the lack of associativity become apparent from a simple example: Let a machine work with a significand of two bits (the leading 1 is not stored). We compute the aforementioned $f = x + y + z$ for $x = 0.875$, $y = -0.75$ and $z = 0.1875$. If we evaluate the formula left to right and normalise the data after each step, we obtain a result of $f = 0.3125$. This is exact. If we however evaluate the second addition first, we obtain $f = 0.375$.

For a machine, the second result is not wrong. It just suffers from an amplification of the individual round-off errors. Indeed, $2x + 2y + z = 0.4375 < 1$ is the "amplification" term of the left-to-right evaluation. Our machine errors are not amplified. For the second variant, we obtain $2y + 2z + x = 1.125 + 0.875 > 1$. Here, round-off errors are amplified.

On a machine, it makes a difference in which order we evaluate the individual terms. Although our round-off formalism yields worst-case estimates, it does give us some hints what we should do in practice: It tells us that some operations are potentially more harmful than others. Multiplying a number by two for example is safe—see our previous discussion on Sect. 4.2.1. Dividing by three might be tricky.

Now, we can dive into the details of our calculation. From the intermediate step (6.2), we see that the error introduced by the first calculation is eventually increased by the next operation.

Reordering calculations

If you can reorder your operations, run the operation introducing the potentially largest error last.

Stagnation

The addition of two numbers is problematic if we add a very small to a very big number. To add them, the ALU has to make them have the same exponent: To add

$1.0|_2 \cdot 2^3 + 1.0|_2 \cdot 2^4$, it first converts the numbers into $0.1|_2 \cdot 2^4 + 1.0|_2 \cdot 2^4$. Throughout this process, we could loose bits: bits are pushed to the right when we denormalise $1.0|_2 \cdot 2^3 \mapsto 0.1|_2 \cdot 2^4$ and thus might be squeezed out of the significand; even if it occupies significant more bits than IEEE precision internally. Therefore, adding a big to a small number—speaking in absolute values—is problematic.

Definition 6.2 (*Accumulation*) If we calculate $f = \sum_{n=0}^{N-1} x_n$ in a register of the ALU, we *accumulate* this result step by step in a computer.

Definition 6.3 (*Stagnation*) Whenever we accumulate, we might run into cases where the partial sum that we have obtained so far is large, whereas the individual numbers we would like to add on top are small. Their contribution towards the whole sum thus is "lost" on the machine. We call this process *stagnation*—the sum should increase, but does not anymore.

Stagnation explains why we have to be careful for example when we compute a scalar product $f = \sum_{n=0}^{N-1} x_n \cdot y_n$ over large vectors where the individual products are of the same order of magnitude.

Numbers of different magnitude

Try to avoid adding or subtracting numbers of different magnitude.

There are ways to work around these machine restrictions. A simple strategy is to sort the entries within an array by their magnitude. That is, adding $1.0 \cdot 2^1 + 1.0 \cdot 2^2 + 1.0 \cdot 2^3 + 1.0 \cdot 2^4 + \cdots$ is less problematic than adding $1.0 \cdot 2^{10} + 1.0 \cdot 2^9 + 1.0 \cdot 2^8 + 1.0 \cdot 2^7 + \cdots$. This doesn't necessarily work if we have to compute a scalar product with two arrays; or it might at least be cumbersome.

In this case, we can employ another trick: We know that we can split up any sum into two sums, accumulate left and right, and then sum up the result: $f = \sum_{n=0}^{N/2-1} x_n \cdot y_n + \sum_{n=N/2}^{N-1} x_n \cdot y_n$. As computer scientist, we can apply this scheme recursively, i.e. split up left and right again. This way, we end up with a tree scheme to obtain the result which is, on the one hand, better-suited to parallelise (we will see this later). On the other hand, it can be more robust if all $x_n \cdot y_n$s are of the same size. We never sum up two partial results which are of completely different magnitude. We never run into stagnation.

Our formalism allows us to formalise this effect:

$$
\begin{aligned}
x_1 +_M x_2 +_M x_3 +_M x_4 &= (x_1(1 + 2\epsilon) + x_2(1 + 2\epsilon) + x_3(1 + \epsilon)) +_M x_4 \\
&= x_1(1 + 3\epsilon) + x_2(1 + 3\epsilon) + x_3(1 + 2\epsilon) + x_4(1 + \epsilon), \text{ yet} \\
(x_1 +_M x_2) +_M (x_3 +_M x_4) &= (x_1 + x_2)(1 + \epsilon) +_M (x_3 + x_4)(1 + \epsilon) \\
&= (x_1 + x_2 + x_3 + x_4) + (x_1 + x_2 + x_3 + x_4)(1 + 2\epsilon)
\end{aligned}
$$

The error impact is more equally distributed among the sum terms: We have $N = 4$ terms in our sum and thus get a maximum error amplification of $(N - 1)\epsilon$ for a plain left-to-right sum. A recursive summation yields a maximum amplification of $\log_2(N)\epsilon$.

Terminology

In different books, you find different terms. Some books call the plain left-to-right summation a "recursive summation" as it is formally equivalent to $f(x_1, x_2, \ldots) = x_1 + f(x_2, \ldots)$. If we split up the range into blocks, compute f over the blocks and then combine the partial results, this is typically called *blocked summation*. The idea to split up a range into blocks, and then to split up these blocks recursively over and over again is called *pairwise summation*, *cascadic summation* or *tree summation*.

Blocked summation will play a major role once we do things in parallel: If we have b blocks, then these b blocks can be summed concurrently. Alternatively, we can sum the first entries of different blocks parallel to the second entries, and so forth. We will use the first pattern on multicore parallel machines and the other pattern on GPUs and vector computers.

Our recursive tree summation uses bipartitioning in the sketch. Nothing stops us from splitting up a range into more sections per recursion step, and nothing stops us to abandon the recursion once blocks become too small. We can construct hybrid schemes.

Datatype

Another flavour of hybrid becomes interesting, when we consider that we have multiple precision formats available on our chips. Recent papers around *fast and accurate blocked summation* not only propose to split up large sums into blocks which fit particularly well to our hardware. They also propose to compute the individual blocks with a lower precision than the total sum. That is, they propose to compute the partial sums for example with single precision, but then to sum up these partial results with double precision. An analysis of such an approach is technical, but the rationale is simple: The lion's share of the work is the summations within the blocks and it thus has to be fast. If the blocks of size b are reasonably small, then our error scales with $(b - 1)\epsilon_{\text{low precision}}$. The flaws in the partial results then are amplified by the summation over the blocks. However, these operations are done with high precision; maybe even with higher precision than what we need in the end by means of our IEEE format of interest. As a result, we have pretty good control over the overall error plus a fast code. This is an appealing example of a *mixed precision* implementation.

Unfortunate constants

We know that our computers work with bits and thus are intrinsically good in working with multiples of two. They however struggle with other values. We furthermore know that many rational numbers introduce a significant round-off error as we have to squeeze them into the significand. $1/2$ is something we can represent exactly, but $1/3$ or $1/10$ (we are not living in a decimal world) are not. As a result, the following two loops yield qualitatively different results:

```
double sum = 0.0;
int N = 10;
for (int n=0; n<N; n++) {
 sum += 0.1;
}
std::cout << sum << std::endl;

sum = 0.0;
for (int n=0; n<N; n++) {
 sum += 0.5;
}
std::cout << sum << std::endl;
```

On my computer, both loops deliver the correct result. Lets increase $N = 10,000$. It still all seems to be correct. If we however switch sum to float, i.e. if we reduce the accumulation variable to single precision, we get 999.903 for the first loop. This is wrong. The errors introduced by each individual rounded 0.1 addition accumulate. The second loop in contrast is fine.

Nice constants

If you work with floating-point constants and if you have some degree of freedom, use "well-behaved" constants such as $1.0 \cdot 2^k$ instead of something that intrinsically introduces errors.

While our programming recommendation is valid, the example is slightly flawed. We said before that we assume that all input data are correct. We might argue that this is not the case in the first loop, i.e. that the increment, the datum, is off a priori. Is it (only) the summation only that goes wrong, or is the wrong output primarily a result of a wrong input? We have to keep this in mind when we argue about the reason for an inaccurate result, but the recipe ("use constants that fit to the IEEE format") is valid no matter what the formally correct explanation is.

Cancellation

Let two values $x \approx y$ be very close. If we compute $x - y$, then the normal-isation on the machine has to introduce new bits behind the decimal point. Let $x = 1.010101 \ldots 0101$ and $y = 1.010101 \ldots 0100$. Then $x - y = 0.000000 \ldots 01$ which we normalise into $1.0 \cdot 2^z$ with the correct z. It is a value close to zero from a machine's perspective. But noone said that this is correct. The machine could have backfilled the result with 1s and given us $1.11111111 \ldots |_2 \cdot 2^z$. Such a result or strategy would by no means be worse. Think about $y - x$ as opposed to $x - y$.

Having two results that are of same "bad" quality, i.e. equally wrong, is not a problem per se. It becomes a problem if we use the result in follow-up calculations. Please note that the present cancellation is something different than error cancel-lation; it actually introduces errors instead of eliminating them! We have seen this effect before: Which values are stored within variables makes a difference. The next rule is not really a new one but rather a specialisation of a previous one:

Postpone cancellation

If your calculations run risk to suffer from cancellation, rearrange the calculations such that the cancellation arises as late as possible.

$(x - y)(x + y) = x^2 - y^2$. Yet, the left side yields bigger errors for $x \approx y$. Also, $zx - zy$ might be better than $z(x - y)$ if x and y are close.

Register assignment and compiler impact

Compilers do a good job of trying to keep all data within the registers as long as possible. Moving data from the register into memory is expensive, so we try to use our full spectrum of fast in-house (register) storage locations. There's a subtlety to this: Registers internally work with higher precision than IEEE, so we reduce the impact of floating point round-offs, too.

It is therefore never a good idea to work on a variable, then not to touch it for ages, and then finally to reuse it. It is always better to do all the work on a variable in one go. Rewrites to achieve this *increase spatial and temporal data access locality*. This is the positive side of things.

However, any instruction reordering can alter the floating point results. If you bring calculations forward, e.g., you might force the compiler to temporary move values out of registers that it otherwise would have kept there. Modern compilers use instruction permutations quite heavily to produce fast code. Whenever we increase a compiler's optimisation level (from -O0 to -O2 (default) to -O3 to ...), we give the translator more freedom to make performance-relevant code modifications. For floating-point numbers, these might not be semantics-preserving.

Most compilers today allow you to disable code permutations: The GNU compiler's switch `-fassociative-math` for example allows the re-association of operands in series of floating-point operations. There's a "don't do this" alternative which explicitly forbids it. Clang supports different floating point policies, too. The default is not "safe".

Compiler impact

Check the compiler's influence on your result from time to time, i.e. try out different optimisation levels, compiler types and generations. In most cases, it makes no difference to the outcome. In some, it does.

Some interesting further material

- David Goldberg: *What Every Computer Scientist Should Know About Floating-Point Arithmetic*. ACM Computing Surveys (CSUR), 23(1), pp. 5–48 (1991) https://doi.org/10.1145/103162.103163
- James H. Wilkinson: *Error analysis of floating-point computation*. Numerische Mathematik. 2, pp. 319–340 (1960) https://doi.org/10.1007/BF01386233
- Pierre Blanchard, Nicholas J. Higham, Theo Mary: *A Class of Fast and Accurate Summation Algorithms*. SIAM J. Sci. Comput., 42(3), A1541–A1557 (2020) https://doi.org/10.1137/19M1257780

Key points and lessons learned

- Arithmetic operations introduce round-off errors, and we have a (worst-case) formalism to study these errors.
- This formalism materialises in few calculation rules.
- In most of the cases, this worst-case analysis is way too pessimistic. However, we have to be prepared.
- We have defined the terms cancellation, loss of significant bits and stagnation.
- Some laws of mathematics (associativity, e.g.) do not "apply" on a computer.
- We have identified a few things one should be aware of as a programmer when we work with floating-point numbers.

SIMD Vector Crunching

7

Abstract

Vector computing can be read as equipping a processing unit with replicated ALUs. To benefit from this hardware concurrency, we have to phrase our calculations as operation sequences over small vectors. We discuss what well-suited loops that can exploit vector units have to look like, we discuss the relation between loops with vectorisation and partial loop unrolling, and we discuss the difference between horizontal and vertical vectorisation. These insights allow us to study realisation flavours of vectorisation on the chip: Are wide vector registers used, or do we introduce hardware threads with lockstepping, how do masking and blending facilitate slight thread divergence, and why does non-continuous data access require us to gather and scatter register content?

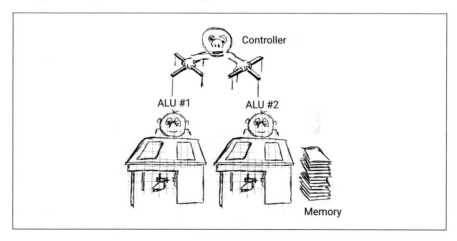

T. Weinzierl, *Principles of Parallel Scientific Computing*, Undergraduate Topics in Computer Science, https://doi.org/10.1007/978-3-030-76194-3_7

Assume we have to add two vectors

$$f_0 = x_0 + y_0$$
$$f_1 = x_1 + y_1$$
$$f_2 = x_2 + y_2$$

$$\cdots$$

just as we did in Chap. 5. Yet, our "machine" is pimped:

- The ALU is told by a controller what to do. The controller acts as master of puppets.
- There are two ALUs and they do exactly the same, i.e. if the controller (master) shouts "add", both add *their* registers.
- The controller is able to address the 2×2 registers separately.

How can we compute y in such an environment way quicker than before? What is the maximum speedup resulting from this architecture? What would happen if we doubled the number of ALUs once more?

A first extension of the simple von Neumann machine (Chap. 5) is the introduction of a vector register: Assume that there is not only a register RegA but that there are two registers RegA1 and RegA2. Logically, we could make register A twice as big such that it can hold (a small vector of) two entries. Alternatively, we could assume that there are two ALUs with exactly the same register types. Both approaches are variations of the same conceptional idea. We obviously assume that the same extensions holds for register RegB, too.

Let there be a new command which we call add2. It takes the content of RegA1, adds RegB1 and stores the result into RegB1 for example. At the same time, the new routine adds RegA2 and RegB2 and stores the result. We call such an addition a *vector operation*, as it effectively runs

$$\begin{pmatrix} y_1 \\ y_2 \end{pmatrix} \leftarrow \begin{pmatrix} x_1 \\ x_2 \end{pmatrix} + \begin{pmatrix} y_1 \\ y_2 \end{pmatrix}$$

in one step.

The potential of such an idea is immediately clear. If we need exactly one cycle for each instruction—an assumption which is wrong, but helps us to bring the message across—then we have to load $2N$ values from the main memory into the chip every time we add two vectors of N elements, i.e. we need $2N$ cycles for all of the loads. We furthermore have to write N values back. The number of computations totals to another N cycles. We end up with $4N$ cycles for the whole vector addition.

With our new vector operation, we reduce the cost for the actual computation by a factor of two. We end up with a total cost of $3N + N/2$ cycles. Not a tremendous saving. Yet, we see where this leads us to: If we now introduce new load and store operations, too, then we can bring down the cost to a half of the original cost. And there is no reason why we should not apply this idea to more than just two registers.

Definition 7.1 (*Vector operation*) A vector operation is an operation which runs the same operation on multiple pieces of data in one go. If programmers or a compiler translate a (scalar) piece of code into a code using these vector operations, they *vectorise* a code.

7.1   SIMD in Flynn's Taxonomy

The idea to introduce vector operations is old. The underlying paradigm has been described by Michael Flynn as *Single Instruction Multiple Data (SIMD)*. The meaning of this is intuitively clear: One single operation (add two values, e.g.) is applied to multiple entries of a vector in one step. Others thus speak of SIMD as *vector computing*.

> **SIMD**

Single instruction multiple data (SIMD) is the hardware paradigm equipping chips with vector computing capabilities.

Historically, vector computing had always been popular with computers constructed for scientific calculations, as our codes traditionally had been written using lots of vector operations—essentially matrix-vector products (the jargon/slang for this is *mat-vecs*). It hence paid off to create specialised hardware. This was in the 1980s. After that, the economy of scale took over. It became cheaper to buy a few more standard computers without these specialised instructions off the shelf. It also was more convenient for developers—no need to replace our additions with specialised SIMD operations, e.g.

Things changed due to the demand for better computer games and movie graphics: If you process an image, very often you apply the same operator to all the pixels. You pipe data subject to the same code through the system. You apply the same operator to one big vector of data. GPU vendors thus started to revive the vector computing idea. One might argue that realistic blood spilling and car races paved the way towards exascale computing with GPUs. As vectorisation is a proper and energy-effective way to provide what users want, it found its ways back into CPUs too. Today, we have SIMD in almost any chip—though with different realisation flavours.

7.2   Suitable Code Snippets

The name vector computing both highlights for which types of codes our paradigm works and it is slightly misleading at the same time. We next try to get a better (formal) understanding which types of code snippets are well-suited for parallelisation:

Let's break down our program into elementary steps as they are supported by the machine, i.e. break down our code into steps like $a \leftarrow b$ and $a \leftarrow b \circ c$ with some operators $\circ$. Each step of the resulting code now becomes a node in a graph. Two nodes n_1 and n_2 are connected via a directed edge from node n_1 to n_2 if the outcome of n_1 is required on the right-hand side of the computation in n_2. The graph of the vector addition $f = x + y$, $f, x, y \in \mathbb{R}^{10}$ corresponding to the code

```
for (int n=0; n<N; n++) {
  f[n] = x[n] * y[n];
}
```

then equals

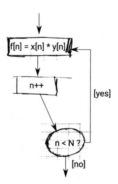

Definition 7.2 (*Control flow graph*) The *control flow graph* is a representation of a particular realisation of our algorithm. It consists of elementary steps connected through causal dependencies.

The control flow graph encodes a partial order: If there's no path in the graph between two nodes n_1 and n_2, we are free to compute either n_1 first or n_2.

Programmers can alter their code by rearranging instructions (nodes) that are independent of each other, and compilers do so all the time when they optimise. Whenever they rearrange/reorder a code, they "only" have to ensure that they do not violate any data dependencies. In the example above, there is nothing we can reorder that fits to our idea of a vector operation. There is however a very simple code transformation:

1. Copy the body within the loop once, i.e. each loop iteration now runs through two entries of the input vectors. In return, the loop increment becomes an increment of two. For the time being, we assume N is an even number.

2. Within the loop body, rearrange the operations such that two operations using the same binary operator but working on different input/output data follow immediately after one another.
3. Finally, fuse these pairs of operations with the same operator into one vector operation.

We obtain the following code:

```
for (int n=0; n<N; n+=2) {
  f[n]   = x[n]   * y[n];
  f[n+1] = x[n+1] * y[n+1];
}
```

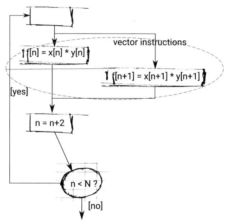

This yields a fundamentally different control flow graph: the update of `y[i+1]` still follows `y[i]` in our textual description. However, the two steps are completely independent of each other. Indeed, the definition of the control flow graph implies that there is no edge between the two multiplications within the loop body. We thus are free to pick any order. Alternatively, we can do both calculations at the same time as one vector operation over tiny vectors of size two. A compiler will immediately recognise this.

> **Loop unrolling**

Loop unrolling by a factor of b is a key technique to generate code exploiting b-fold vector units.

It is clear what vector operations within our (unrolled) loop body have to look like:

- They have to group exactly the same operation together (after all, it is called "Single Instruction"),
- each operand has to have the same type (don't mix `double`s and `int`), and
- the results have to go into separate memory locations.

Compared to our vector addition, it is more challenging to bring the inner product $f = \sum_{n=0}^{10-1} x_n \cdot y_n$ onto a vector machine: we can do the multiplications in parallel, but the accumulation of the result—so far—has to be done step-by-step.

Definition 7.3 Let $f(x_1, x_2)$, $f : \mathbb{R}^2 \mapsto \mathbb{R}$ be a commutative and associative binary operator. The computation of $r = f(f(\ldots f(f(x_1, x_2), x_3), \ldots), x_N)$ is a *reduction*.

Definition 7.4 (*Collective*) A collective operation is an operation which involves all parallel compute entities.

In most cases, the ∘ above is an addition or multiplication. The need for a reduction to realise the scalar product limits our gain from vector operations. However, nothing stops us from reusing our blocking idea from the floating point discussion:

```
double f = 0.0; double tempF1 = 0.0; double tempF2 = 0.0;
for (int n=0; n<N; n+=2 ) {
  tempF1 += x[n] * y[n]; tempF2 += x[n+1] * y[n+1];
}
f = tempF1 + tempF2;
```

With this pseudo code on our desk, we have a control flow graph fit for a SIMD machine. Here, operation pairs can be replaced by one vector operation. We will make compilers to do this job for us later. They will yield significantly improved performance "automatically".

So far, we have unrolled the loop once and assumed that the loop cardinality is even. For $N = 10$, this holds. If we have a vectorisation parallelism of a factor of four—think about four-tuples of registers—we have to rethink our solution. A straightforward solution in this case would be to augment the input vectors x and y by two zeroes each, i.e. to make them $\hat{x}, \hat{y} \in \mathbb{R}^{12}$. This is not always possible or desirable—we cannot always extend data structures in memory without overwriting information and if we rewrite code, we might be hesitant to sacrifice the memory.

What we can do instead is to split the vector logically into two parts:

$$x = (\underbrace{x_0, x_1, \ldots, x_7}_{\text{first part}}, \underbrace{x_8, x_9}_{\text{second part}}).$$

We now can apply the code transformation pattern to the first part with a vectorisation factor of four, and then handle the remaining two entries of the input vectors with another "loop" (it is a degenerated one with only one run-through) and an unrolling factor of two. Again, we will later throughout this book rely on a compiler to do this for us. But we have to understand these patterns to understand why certain codes yield a certain performance.

> **Loops are the workhorses of vector computing**

Vectorisation is a natural fit to loops. The other way round, loops are the workhorses behind most efficiently vectorised code. To make a loop a candidate for vectorisation, it has to be in canonical form.

Definition 7.5 (*Loops in canonical form*)

1. *Countable.* We have to know how many loop iterations are to be performed when we first hit the loop. A while loop where we check after each sweep whether a criterion holds is not a fit for vectorisation.
2. *Well-formed.* Some people like loops with break or continue statements or loops which can jump out of a function. More "sophisticated" programmers make loop bodies throw exceptions or alter the loop counter within the loop body. All of these are poison to vectorisation. They make it impossible for the compiler to translate loop instructions into vector instructions.
3. *Branching/simplicity.* The loop body has to be simple.

   a. Loops should not host if statements. The idea behind all vector computations is that you do the same thing for all entries of a vector. If one vector entry goes down one source code path and the others do something different, this contradicts the idea behind SIMD. We will weaken this statement on ifs in a minute, but in principle we would like to have a no-if policy.
   b. Vectorisation can only work for inner loops. If we have nested loops, we can't expect the machine to vectorise over both the outer loops and the inner loop; at least not if the inner loop's range depends on the outer loop indices.
   c. We cannot expect vectorisation to work if we use function calls (think about something like trigonometric functions in the worst case) within the loop.

On top of the "hard" constraints formalised by the term canonical form, loops have to exhibit a reasonable *cost*. Vectorisation is not for free. There are multiple technical reasons for this: Most chips for example downclock once they encounter vector operations. They become slower. If you get it right, this downclocking does not eat up all of the speedup of vectorisation. But it reduces the pay off. In the worst case, it however renders vectorisation counterproductive.

We will discuss in detail later how to force or encourage a compiler to vectorise. In most cases (and for most programmers), code that can be vectorised nicely is code written in a way such that a compiler can automatically create the right vector instructions for the code fragment. Such a code follows the recipes above.

7.3 Hardware Realisation Flavours

There are two fundamentally different ways to realise SIMD in hardware: We can work with large registers that host N values at once. When we add two of these massive registers, they effectively perform N additions in one rush. Alternatively, we can work with $2N$ normal registers where the N pairs of registers all perform the same operation.

7.3.1 Large Vector Registers

The former realisation variant is what we find in standard processors today. Though there are numerous early adoptions of the SIMD concept, today's architectural blueprint dates back to around 1999 when Intel introduced a technique they called SSE. The main processor here is sidelined by an FPU (floating point unit as compared to ALU) hosting additional registers. These registers are called xmm$k$ with $k \in \{0, 1, \ldots, 7\}$. They are larger than their ALU's counterparts. In the original SSE, they had 128 bits. SSE can only be used for single precision arithmetics—its primary market had been computer games and graphics—which means each xmm register can host up to four single precision values with 32 bits each. If four entries of a vector $x \in \mathbb{R}^4$ are held in xmm0 and four entries $y \in \mathbb{R}^4$ in xmm1, then the addition of xmm0 and xmm1 computes four additions in one rush. The hardware ensures that xmm0 does not spoil xmm2 and so forth. The xmm$k$ registers are the RegA, RegB, …registers from our introductory example, i.e. the RegA1, RegA2, and so forth are physically stored in one large RegA register.

ALU versus FPU

If we use standard binary and unary floating point operations, the computation will be done by the ALU directly within the CPU. If we use vector operations, the FPU will take over which sits next to the CPU. Although they are typically tightly integrated, the FPU can run with reduced frequency, and it is slightly slower in accessing memory.

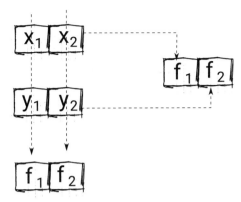

Fig. 7.1 Let $x = (x_1, x_2)$ and $y = (y_1, y_2)$ be two vector registers holding two values each. We can combine these registers *vertically* into $f = (x_1 \circ y_1, x_2 \circ y_2)$ or *horizontally* into $f = (x_1 \circ x_2, y_1 \circ y_2)$

Single precision support is not what we need in science. With SSE2, Intel therefore allows us to store two double precision rather than four single precision values per register. They also offer many new operations and support for further (integer) data types, but the single and double precision are the predominant use cases for scientific computing. With SSE3, new instructions were made available which allow us to add two (or four) values stored in one register. Before that, we had been able to add xmm0 and xmm1 but not the four values within xmm0, e.g.

Definition 7.6 (*Horizontal versus vertical vector operations*) If one vector register holds two entries and if we combine its first and second entry, this is a *horizontal vector operation*. If entries from multiple registers are combined with each other, we call this a *vertical operation*.

Horizontal operations work with tuples of vector registers (Fig. 7.1). Each entry in the outcome vector results from the combination of two neighbouring entries from one source vector register. Vertical operations combine the entries from different vector registers. The operations introduced as RegA1, RegA2, …combined with RegB1, RegB2, …in this chapter are vertical SIMD operations. If we added RegA1 with RegA2, we would have introduced a vertical vector operation. At first glance, it is not obvious why one should introduce horizontal operations; notably as vertical operations are way faster than their horizontal "counterparts". However, they are convenient whenever partial results are already distributed among entries of one vector register. Without horizontal operations, we would have to disentangle a vector register before we can combine its individual entries.

Vertical versus horizontal vector operations

We compute $f = \sum_{i=1}^{2} x_i y_i$, i.e. a small vector product, with a vector length of two. Our code loads (x_1, x_2) into the first register, (y_1, y_2) into the second, and then multiplies them component-wisely via one vertical operation. Thus, there will be one vector register holding $(x_1 y_1, x_2 y_2)$. Without horizontal vector operations, we next have to decompose (split) this vector register up into two registers—another step—before we eventually add up the partial results.

Further improvement of vector computing capabilities results from the fact that modern vector units offer fused multiply add (FMA): They compute $f = x + (y \cdot z)$ in one step. That is two arithmetic operations (a multiplication plus an addition) in one step rather than two! The operations are fused.

Beyond the extensions of the vector instruction set, the biggest improvement upon SSE is SSE's successor Advanced Vector Extensions (AVX), which widens the individual register from 128 bits to 256. Later, we got the AVX-512 extension. Eight double values a eight bytes now fit into one register.

Performance breakdowns

Statements on the pay off of vector operations as factors of two or four lack two details: On the one hand, vector operations typically have a way higher latency than their scalar counterparts. That means, loading data into vector registers is expensive and we have to amortise this speed penalty by vector efficiency. On the other hand, vector units are independent of the CPU. Vendors thus drive them with slightly different clock speed. They reduce the frequency for AVX-heavy code.[1] Otherwise, the chip would become too hot. We conclude that optimal code, from a vector point of view, relies on sequences of $f = x + (y \cdot z)$ operations, but the impact on the time-to-solution has to be analysed carefully and experimentally.

7.3.2 Lockstepping

An alternative to large vector registers is the use of many standard registers that all do the same thing at the same time. They synchronise after each step. The selling point compared to scalar architectures remains the same: We can save hardware and energy for all the orchestration and execution logic.

Definition 7.7 (*Lockstepping*) If multiple ALUs do exactly the same thing at the same time, they run in *lockstep mode*.

---

[1] In return, vendors sell you a temporary increase of the clock speed as turbo boost; which sounds nice but actually means that you will not benefit from this with a floating-point heavy code.

GPUs traditionally rely on lockstepping. NVIDIA's CUDA model[2] speaks of multiple threads per processor. Yet, these threads are no threads in the CPU sense. They refer to massive SIMDsation through lockstepping, i.e. they are lightweight threads compared to what we call thread on the CPU.

Definition 7.8 (*Single Instruction Multiple Threads*) On accelerators (GPGPUs), SIMD programming is called *single instruction, multiple threads* (SIMT). Several (lightweight) threads run the same instruction but on different pieces of data.

We summarise that SIMT is one way to realise the vision behind SIMD. The other way is the large vector registers found in CPUs.

7.3.3 Branching

Both vector units and SIMT architectures today offer some support for branching. They can evaluate codes like

```
for (int n=0; n<N; n++) {
  if (x[n]>0.0) {
    f[n] = x[n] * y[n];
  }
  else {
    f[n] = -x[n] * y[n];
  }
}
```

At first glance, this comes as a surprise, as an input to this code snippet might be $x = (1, -1, 1, -1, \ldots)^T$. There seems to be no real potential for vectorisation. However, modern systems use a technique called masking:

Definition 7.9 (*Masking*) *Masking* means that a condition is evaluated and its outcome is stored in a register. A bitfield with one entry per vector register entry is sufficient. Once a SIMD computation is completed, the result is discarded for those vector elements for which the masking's register bit is not set.

---

[2] Two different things are called CUDA: There is a programming language (a C dialect) that we call CUDA, and there is a hardware architecture paradigm called CUDA. We do not discuss the CUDA programming here.

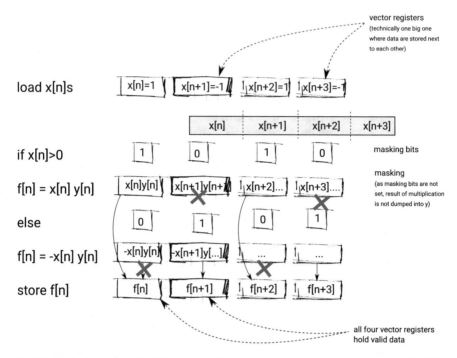

load x[n]s

if x[n]>0

f[n] = x[n] y[n]

else

f[n] = -x[n] y[n]

store f[n]

Fig. 7.2 Illustration of masking/if statements of our example code snippet with a vector multiplicity of four. In the if/else branches, we only use 50% of our compute resources as the impact of every second vector calculation is masked out. This is also called thread or vector divergence

If we have 4 (GPU) threads or a vector register hosting 4 entries, masking can deactivate the impact of any of the 4 ALUs independently. In principle, they all compute the same. We just ignore the outcome if our masking kicks in (Fig. 7.2).

Definition 7.10 (*Divergence*) If k threads/ALUs compute the same in lockstepping mode, yet some of the results will be masked, i.e. some computations are unnecessary, the threads *diverge*.

Divergence is a direct consequence of masking: Multiple threads compute, but only a subset of them contribute towards the calculation. Some people prefer the term *control divergence*. It effectively sacrifices scalability potential. Even worse, we run calculations and pay the energy price for them, even though these results are not required eventually. Masking therefore is a double-edged sword: It allows us to vectorise over code snippets with if statements, but it can lead to code that wastes energy and is actually slower than its scalar counterpart, as vector units run with reduced frequency and have higher memory access latency.

Particular severe is the impact of masking for if-else statements as used above: Over the two branches, we will always get a divergence of around 50%. Different to the classic scalar branching, masking and branching in a lockstepping code can never skip any part of the code. We always execute everything line by line. The masking solely can mask out the effect of complete if or else blocks. The problem can be mitigated—though only from a performance point of view, not from a "wasted" computation point of view—via blending:

Definition 7.11 (*Blending*) If a code has two branches, *blending* implies that we evaluate both branches simultaneously but only use one of the outcomes.

Blending can be read as natural extension of masking: We don't mask out but make the masking bit decide which result to use in the end. This gives a compiler a larger set of options how to arrange computations (and how to vectorise them).

Branches in loops

Our requirements for loops on Sect. 7.2 are too strict: As we have masking and blending, we can use conditionals, but (i) they have to be simple—complete if-then-else cascades or branches in outer loops do not work—and (ii) they make our code less efficient.

It is unlikely that we will go beyond 512-bits versions of vector instructions on x86 machines any time soon, as all caches and the memory are organised in 512-bit blocks: Your main memory is a big sequence of storage locations that you can address individually, but the machine internally works with blocks of 512 bits. This is similar to a book where you can look for individual paragraphs, but you can browse through the book only page-wisely.

You cannot load any four consecutive doubles into one vector register in one rush. You can only do so if they are stored within the same block (aka book page; otherwise you have to flip pages while you load). Therefore, vector operations unfold their full potential whenever we can load and store full registers, i.e. multiple vector entries, in one rush.

Whenever entries x_0, x_1, x_2, …of a vector x are stored consecutively and at the right place in memory, the computer can take a block of k x_n values and load them into the one large vector registers in one step. "At the right place" means that the first entry x_0 should be placed at an address which coincides with the start of a 512 bit block. It should be *aligned*. Otherwise, we speak of an *unaligned* memory access. Traditionally, vector loads and stores have always been expensive whenever the entries of a vector have not been consecutively stored in memory and/or not aligned:

Definition 7.12 (*Gather and scatter*) If we have to grab various entries from different places in memory into one large vector register, we *gather* information. The counterpart is called *scatter*. They are *collective* operations, as they affect all vector register entries.

In the SSE era, gathers and scatters have been so expensive that it often was disadvantageous to use vector operations at all for scattered data. This changes. The reason why scatter and gather are so important that vendors start to support them better in hardware becomes clear when we revisit matrix operations:

Matrix-matrix multiplication

If you have a matrix, you can either store it row by row (C-style) or column by column. The latter is the Fortran way. When you multiply two matrices $X \cdot Y$, you can either take the first row of X times the first column of Y, then the first row or X and the second column of Y, and so forth. Or you can take the first column of X and multiply it with the first row of Y. The result is a matrix to which you next add the result of the second column of X times the first row of Y. No matter which variant you follow and no matter whether you store your matrices in row-major order (this is the C way) or column-major (Fortran), one of your matrix accesses always will suffer from scattering.

A new trend around vectorisation—in particular on the GPU side and in particular focused on machine learning—is the progression from vector operations to matrix calculations. Classic vectorisation orbits around operations with small vectors. Due to the advance of machine learning, calculations with small matrices become more and more important. With out-of-the-box SIMD, we have to break matrix operations down into sequences of operations over vectors. New machines however offer so-called *tensor units*:

Definition 7.13 (*Tensor units*) Tensor units are dedicated hardware to compute $F = X + (Y \cdot Z)$ and variations of these, where F, X, Y and Z are (tiny) matrices. They translate the vectorisation idea into the matrix world. Some vendors speak of *tensor cores* as they are separate cores to the standard processors on the GPGPU.

With tensor units we witness a little bit of history repeating: Again, they became first available for reduced (half) precision, before they have grown into hardware that supports IEEE precisions as used in most scientific codes, i.e. single and double precision.

Some material to dive into details

- Dominic E. Charrier, Benjamin Hazelwood, and others: *Studies on the energy and deep memory behaviour of a cache-oblivious, task-based hyperbolic PDE solver.* The International Journal of High Performance Computing Applications. 33(5), pp. 973–986 (2019) http://doi.org/10.1177/1094342019842645
- Johannes Hofmann, Jan Treibig, Georg Hager, Gerhard Wellein: *Comparing the performance of different x86 SIMD instruction sets for a medical imaging appli-*

cation on modern multi- and manycore chips. WPMVP '14: Proceedings of the 2014 Workshop on Programming models for SIMD/Vector processing, pp. 57–64 (2014) https://doi.org/10.1145/2568058.2568068

> **Key points and lessons learned**

- We know what a vector computer is and that it belongs to the class of SIMD architectures.
- Control graphs are one tool to analyse given code and to identify its vectorisation potential.
- We have defined the term reduction and identified implications for the arising vector efficiency.
- There are four criteria loops have to meet to benefit from vectorisation.
- Different vendors use different terminology and technical concepts to realise SIMD—in particular GPUs versus x86 chips. The vendor strategies mainly differ in whether they use large registers to hold multiple vector entries or rely on lockstepping. The implications for programmers however remain similar. If you can write CPU code that vectorises brilliantly, you can usually transfer this code to GPUs relatively easy.
- Masking and blending are the two key concepts how vector computing facilities support branching. They imply that we can use if statements in the inner loop but also imply that we should be careful when we use them as they might harm the performance. The latter insight is formalised by the term thread divergence.

Arithmetic Stability of an Implementation

8

Abstract

Round-off errors are tied to the implementation, i.e. two different implementations of the same algorithm might exhibit different error propagation patterns. We introduce the term arithmetic stability and formalise how to find out if an implementation is stable under round-off errors. The formalism allows us to show that our previously studied N-body problem is indeed single-step stable and also stable over longer simulation runs. For implementations of extreme-scale scalar products, e.g., stability however does not come for free, and we have to carefully arrange all computational steps.

We compute

$$f(x_1, x_2) = x_1 - x_2 = \frac{x_1^2 - x_2^2}{x_1 + x_2}$$

for $x_2 = 1$. If we plot either the left or the right variant of the above equation, they seem to compute the same. However, if we compare the right to the left side, we observe that they do not yield the same result for all x_1:

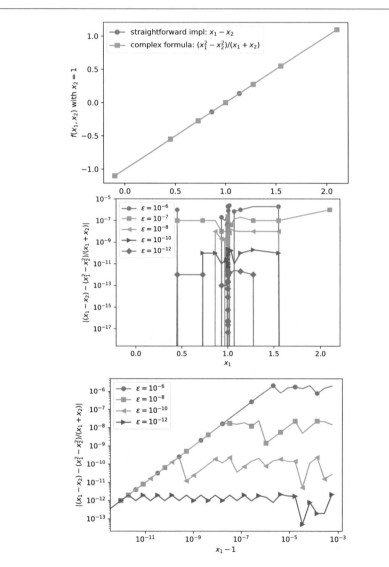

The different plots are for different precisions, i.e. we set the machine precision to 10^{-6}, 10^{-7}, 10^{-8} and 10^{-10}. The higher the precision, the smaller the difference between the left and the right variant. For reasonably big differences between x_1 and x_2, both sides of the equation yield the same value. For small differences, the two sides however yield significantly different results. Which variant (left or right) gives the more accurate one, and how can we show this? What does this mean for larger calculations where $x_1 - x_2$ is only an intermediate step?

The left and right side of

$$\cdot \quad f(x_1, x_2) = x_1 - x_2 = \frac{x_1^2 - x_2^2}{x_1 + x_2}$$

are (symbolically) exactly the same. Otherwise, the equality sign would be ill-placed. However, the left and right side yield different results on a computer. The right variant of the formula consists of way more operations. It is more expensive and concatenates more round-off errors. Once we plug in our machine calculation rules—remember that we study x_1 close to x_2—we however see that

$$x_1 -_M x_2 = \epsilon \quad \text{but} \tag{8.1}$$

$$\left(\frac{x_1^{2M} -_M x_2^{2M}}{x_1 +_M x_2} \right)_M = \left(\frac{(x_1^2 -_M x_2^2)(1 + \epsilon)}{(x_1 + x_2)(1 + \epsilon)} \right)(1 + \epsilon) \tag{8.2}$$

$$= \left(\frac{(x_1^2 -_M x_2^2)(1 + \epsilon)}{(x_1 + x_2)(1 + \epsilon)} \right)(1 + \epsilon)$$

$$= \left(\frac{x_1^2 -_M x_2^2}{x_1 + x_2} \right)(1 + 4\epsilon) \tag{8.3}$$

$$= \left(\frac{\epsilon}{x_1 + x_2} \right)(1 + 4\epsilon)$$

The analysis of the straightforward implementation of the difference in (8.1) is, well, straightforward. The analysis of the alternative formulation (8.2) is more interesting: In a first step, we mechanically insert our replacement rules. We then drop all higher order terms (see the remarks in Sect. 6.3) and therefore also get rid of the division by $1 + \epsilon$, before we finally once again rush into cancellation when we compute $x_1^2 - x_2^2$ in (8.3).

In our "sophisticated" rewrite, we obtain a factor $(1 + C\epsilon)$ with $C = 4$ that we multiply with the solution. This factor is bigger than the sole ϵ that we get in (8.1). However, we see that the "large" error term is scaled with $(x_1 + x_2)^{-1}$. That suggests that the complex rephrasing of $x_1 - x_2$ is of value for very close variables x_1, x_2 as soon as $|x_1 + x_2|$ is sufficiently large. We have to pay for the additional computations, but the error is damped. Otherwise, just writing down $x_1 - x_2$ is the method of choice.

8.1 Arithmetic Stability

Definition 8.1 (*Absolute and relative error*) Let $f_M(x)$ be the outcome of an implementation of $f(x)$ evaluated on a machine. $f_M(x)$ is subject to round-off errors. The *absolute error* is $e = f_M(x) - f(x)$. The *relative error* is

$$\frac{e}{f(x)} = \frac{f_M(x) - f(x)}{f(x)} = \frac{f_M(x)}{f(x)} - 1.$$

This is an equivalent of the Definition 4.1 with the exact outcome of a function as baseline.

Three remarks

1. Most people, books and papers put absolute brackets around the error formulae. They ensure that errors are always positive. Sometimes, it is however useful to highlight whether an error means "too much" or "not enough".
2. For $f(x) \mapsto 0$, the definition introduces a division by zero. In this case, you might have to use L'Hôpital's rule.
3. If you have $f(x, y)$, you have to specify an error w.r.t. x and one w.r.t. y. They can be qualitatively different.

A piece of code as we write it down can be read as a sequence of mappings (functions). Some of them are unary, i.e. accept only one argument; the x^2 in the introductory example, e.g. Others are binary. All map from one set or two sets into another. An expression like $\left(\frac{x_1^2 - x_2^2}{x_1 + x_2} \right)$ can be written down as long sequence of functions $f^{(N)} \circ \ldots \circ f^{(2)} \circ f^{(1)}$ applied to our input data. The superscripts $^{(k)}$ here just enumerate the different fs in a row. Such expressions are read from right to left, i.e. we apply $f^{(1)}(x)$, pipe the result into $f^{(2)}$, and so forth.

With each function, we are slightly off. The "off"s accumulate as we evaluate $f_M^{(N)} \circ \ldots \circ f_M^{(2)} \circ f_M^{(1)}$ (Fig. 8.1). We are interested how much we are off and whether the error becomes bigger and bigger or remains under control:

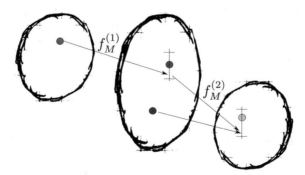

Fig. 8.1 Schematic illustration of the evaluation of $y + x^2$, e.g.: A first function $f_M^{(1)}$ computes the square, but the result is slightly off the analytical value due to round-offs. This is a unary function. The slightly skewed result is used by $f_M^{(2)}$ which computes the sum of two values and again yields a slightly wrong result within an area around the real result (cmp. error bar around real result denoted as point)

Definition 8.2 (*Arithmetic stability*) An algorithm is *arithmetically stable* if its accumulation of round-off errors does not explode, i.e. if there is a fixed C such that the absolute relative error is bounded by $C\epsilon$.

Stability analysis assumes correct input. It thus characterises to which degree implementations give the "(slightly) wrong answer to the correct question". Unfortunately, the definition of arithmetic stability is not the same in all books. Some differ quite dramatically. Many maths books for example call an algorithm stable if

$$\frac{|f_M(x) - f(x_M)|}{|f(x_M)|} \le C_1\epsilon \quad \text{for} \quad \frac{|x_M - x|}{|x|} \le C_2\epsilon. \tag{8.4}$$

I wrote before that I usually use C as generic constant. Along these lines, I could have used C twice instead of the C_1 and C_2. However, I wanted to emphasise here that the two Cs differ from each other. The right-hand side of (8.4) states that an exact (correct) input x has to be an idealisation. On computers, x is subject to machine precision/truncation effects (or other "flaws") as well, and we thus have to deal with a machine approximation x_M. As long as this difference remains bounded, a stable algorithm yields "almost the right answer to almost the right question". This is the statement behind the left inequality in (8.4). Equation (8.4) is more sophisticated than our initial notion of stability as it accommodates deviations in the input data. The less formal, hands-on characterisation of stability is sufficient for us.

More important than the philosophical question whether you can represent input accurately or not is the fact that you will find books that don't even speak of "stable" but rather prefer statements along the line "one implementation is more stable than the other one". This is a convenient way to phrase things: There's a set of problems that a computer can solve accurately. We still have to formalise this set. Within the set, there are codes that give the right answer and there are codes, i.e. implementations, that don't give the right answer—even though they all implement (symbolically) the same thing. Some implementations run into rounding issues while others do not. Between both extremes, there are codes that yield accurate results for some input (Fig. 8.2).

Arithmetic stability analysis

When you study the stability of an implementation w.r.t. round-off errors, you have to break down the write-up of the code into elementary operations.

Our notion of arithmetic stability takes the round-off error analysis and relates it to the real outcome of a piece of code. We ask whether the outcome is *accurate*. In the example, we see that

$$\epsilon \cdot \frac{1 + 4\epsilon}{x_1 + x_2} \ll \epsilon$$

if $x_1 + x_2 \gg 1 + 4\epsilon$. There is a threshold which determines which of the two implementation variants is more stable (read "accurate"). Further to these accuracy

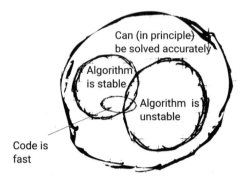

Fig. 8.2 Schematic illustration of problem/code classes: Among the problems we want to solve in scientific computing, we are interested in those that we can solve accurately. This property is one we still have to formalise. Within this category, there are implementations that are always stable w.r.t. round-off errors and some that are never stable. In-between, there is a blurry region, i.e. most codes are stable for some inputs and not for others. Among stable, unstable and in-between codes, some implementations are fast—they exploit features such as vector units and have low computation counts, e.g.—others are not

statements, our analysis demonstrates that neither variant is stable always: A simple limit where we make x_2 approach x_1 shows

$$\lim_{x_2 \mapsto x_1, x_2 \neq x_1} \frac{(x_1 -_M x_2) - (x_1 - x_2)}{x_1 - x_2} = \frac{\epsilon - x_1 + x_2}{x_1 - x_2} = \infty$$

$$\lim_{x_2 \mapsto x_1, x_2 \neq x_1} \left(\frac{x_1^{2M} -_M x_2^{2M}}{x_1 +_M x_2} - (x_1 - x_2) \right)_M \cdot (x_1 - x_2)^{-1} = \infty.$$

In our analysis, we can arbitrarily increase the relative error for both implementation variants. On a machine, it is impossible to choose x_1 and x_2 arbitrarily close, i.e. there is a natural upper bound on the absolute error. But it can be very high. A high error might, in more complex code, feed into follow-up calculations, and that is where the problems start.

Definition 8.3 (*Loss of significance (again)*) If the *absolute relative error increases faster than the absolute error*, we basically loose all the bits in our floating point number that carry meaningful information.

Cancellation is the extreme case of a loss of significance. In general, any arithmetically unstable implementation suffers from this loss. It means that the outcome of the algorithm runs risk to be totally meaningless. However, an unstable implementation in a numerical sense does not necessarily lead to a loss of significance. Everything we do here is a worst-case analysis. As errors cancel out, we might actually be lucky.

Garbage in, garbage out

Computer scientists love the phrase "garbage in, garbage out". Our insight clarifies that even meaningful input can lead to "garbage out" due to round-off effects.

8.2 Stability of Some Example Problems

8.2.1 The N-Body Model Problem

We return to our model problem from Chap. 3 and ask the question whether the algorithm proposed is vulnerable to round-off errors. For this, we abstract the problem and study

$$p(t + dt) = p(t) + dt \cdot F(p(t)). \tag{8.5}$$

We argue over a $p(t)$, and assume that there is a "magic" function F which updates this $p(t)$.

Applying the analysis to the example code

Assume that the p has two entries per particle, i.e. is a vector with the actual particle position plus its velocity. F then returns two values, too. For the first one, it simply returns the particle velocity in our code, while its second result part encodes the weighted sum over the forces. If you keep this vector or tuple interpretation in mind, the analysis fits component-wisely to the example code.

As we plug in our machine arithmetics into (8.5), our stability analysis accepts that the time evolution is subject to machine errors. Before we start, we however make one more assumption: We assume that the calculation of $F(p(t))$ is stable. If F were not stable, it would not make any sense to continue to investigate the stability of the overall code. Hence, let $F_M(p(t)) = F(p(t))(1 + C\epsilon)$, and we obtain

$$
\begin{aligned}
p_M(t + dt) &= p(t) +_M dt \cdot_M F_M(p(t)) \\
&= p(t) +_M dt \cdot_M F(p(t)) \cdot (1 + C\epsilon) \\
&= p(t) +_M (dt \cdot F(p(t)) \cdot (1 + C\epsilon)) (1 + \epsilon) \\
&= p(t) +_M (dt \cdot F(p(t))) (1 + (C + 1)\epsilon) \\
&= p(t) +_M \left(dt \cdot F(p(t)) + \underbrace{(C + 1)\epsilon dt \cdot F(p(t))}_{\text{small}} \right) \\
&= p(t) +_M dt \cdot F(p(t)) \\
&= (p(t) + dt \cdot F(p(t))) (1 + \epsilon) \\
&= p(t) + dt \cdot F(p(t)) + \epsilon p(t) + \underbrace{\epsilon dt \cdot F(p(t))}_{\text{small}} \\
&= p(t) + dt \cdot F(p(t)) + \epsilon p(t).
\end{aligned}
$$

The derivation above uses some explicit assumptions:

1. We still assume that $p(t)$ is correct. Therefore, I use a subscript M for $p_M(t+dt)$, but I use no subscript for $p(t)$.
2. We assume that F's computation is stable (as discussed above).
3. We assume that we have a small time step size dt. Twice, we thus kick out terms scaled with $\epsilon \cdot dt$.

The value we start from, i.e. $p(t)$, scales the impact of the round-off error. We can consider $p(t)$ to be a constant handed into the (first) time step computation, as it is given from the outset.

> **One-step round-off stability**

Our model problem code is one-step stable.

The stability statement refers to one step, as we exploit that we consider $p(0)$ to be correct. Some people hence call it *local stability*. After one step only, we however have a flawed $p_M(dt)$. This flaw propagates through subsequent iterations. It is hence not valid to claim "one step is stable, therefore a sequence of steps is stable, too". The one step stability relies on a correct (in a rounding sense) input and this assumption is already wrong for step two.

When we make a statement on the whole algorithm, we thus have to take the number of steps into account and can only assume something on the initial condition $p(0)$. To determine the error after N time steps, we can derive the following two equations following the argument from above:

$$p_M(0) = p(0) \quad \text{is exact, but}$$
$$p_M(N \cdot dt) = p_M((N-1)dt) + dt \cdot F(p_M((N-1)dt)) + \epsilon p_M((N-1)dt)$$
$$= p_M((N-1)dt)(1+\epsilon) + dt \cdot F(p_M((N-1)dt)).$$

This is an induction formula. We unfold it and end up with

$$p_M(N \cdot dt) = p_M((N-1)dt)(1+\epsilon) + dt \cdot F(p_M((N-1)dt))$$
$$= p_M(0)(1+\epsilon)^N + \sum_{n=0}^{N-1} dt \cdot F(p_M(n \cdot dt))$$
$$= p(0)(1+N\epsilon) + dt \cdot \sum_{n=0}^{N-1} F(p_M(n \cdot dt))$$
$$= p(0)(1+N\epsilon) + dt \cdot C \cdot \sum_{n=0}^{N-1} F(p(n \cdot dt)),$$

where we, once again, assume that F does not yield a dramatically different output if we plug in slightly skewed input. In this formula, the very first value, i.e. our initial value, times the number of time steps scales the error. However, the error does not explode. It grows linearly in the number of time steps.

For a given time interval from 0 to T, our code cuts the continuous time into pieces. We have

$$N \sim \frac{T}{dt}.$$

Once we fix the observation interval $(0, T)$, we know the number of time steps N that we need to step through time. There is a constant incorporating N such that the growth of the relative error is bounded.

> **> Global round-off stability**

Our model problem (a so-called explicit Euler) is stable w.r.t. round-off errors.

8.2.2 Logistic Growth

We next study logistic growth which describes a scalar function $p : t \mapsto \mathbb{R}$. Our code realises the time stepping scheme

$$p(t + dt) = p(t) + dt \cdot \left(a \cdot p(t) \cdot (1 - \frac{p(t)}{b}) \right),$$

i.e. we know that it is stable as long as $F(p) = a \cdot p(t) \cdot (1 - \frac{p(t)}{b})$ is stable. A valid assumption.

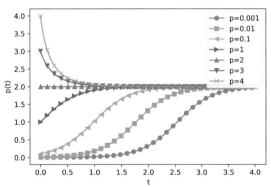

The analytical solution models the population of a species $p(t)$. The species replicates and dies. The higher the population count p the more children are born, i.e. the faster p grows. This replication rate is calibrated through the parameter a. However, if the species' population exceeds a magic threshold introduced by the finite amount of food, members of the population starve. The constant b in the model parameterises this threshold. The above plot shows typical solution curves for different initial p (population) values.

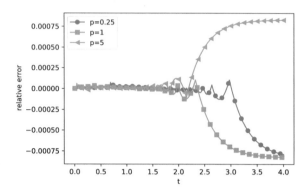

In the plot above, I run each simulation per initial $p(0)$ value twice. Once, I select 60 valid digits in decimal precision. This is ridiculously high, so we can assume that there is basically no arithmetic error. In the other case, I use only six valid decimal entries in the significant. I track the relative difference

$$\frac{p(t)|_{6 \text{ digits}} - p(t)|_{60 \text{ digits}}}{\max(p(t)|_{6 \text{ digits}}, p(t)|_{60 \text{ digits}})}.$$

over time. Even though all three experiments converge towards the same value of $p(t) = 2$ over time, we obtain qualitatively different curves for the relative differences.

For $t \ll 2$, all three studies are subject to rounding errors, but these cancel out. The errors wobble around, yet stay close to zero. This confirms our statement that our stability analysis plays through worst-case scenarios: In this regime, the rounding error sometimes overshoots, sometimes yields something slightly too small, and effectively these effects cancel out.

Around $t \approx 2$, the wobbling amplitudes grow significantly and the rounding errors start to accumulate. They do not cancel out over multiple time steps anymore. This happens the earlier the bigger $p(0)$: The initial value scales the global round-off error as it builds up and with larger $p(0)$ we hit the threshold after which errors start to accumulate earlier. This is the worst-case regime we are afraid of: Each time step adds yet another error, they all have the same sign, and we thus get a worse result over time.

Definition 8.4 (*Round-off error propagation*) For time-stepping codes, round-off errors *propagate through time*, i.e. a round-off error at time t induces an error for $\hat{t} > t$.

All differences are bounded for $t \gg 4$, i.e. on the long term the choice of $p(0)$ seems not to make a difference. This does not come as a surprise, as the updates to $p(t)$ in the time stepping algorithm all go to zero as $p(t)$ approaches $p(t) = 2$ and the solution becomes stationary. We study this behaviour later under the rubric *attractive solutions* (Sect. 12.2). For the time being, we ignore this and observe: Once a solution has approximated $t > 4 \pm 0.0008$, the right-hand side evaluation seems to yield zero, as the numerical approximation $p_M(t)$ does not change anymore. The ± 0.0008 is the effective machine precision for this calculation with our artificial machine with six valid decimal digits.

8.2.3 Scalar Product

The scalar product is a fundamental building block of many scientific applications. When we throw our forward analysis rules onto an implementation of the scalar product via a for-loop, we get a pattern that is the prototype of an induction proof:

$$x_0 \cdot_M y_0 = (x_0 y_0)(1 + \epsilon), \quad \text{and}$$

$$\left(\sum_{n=0}^{N-1} x_i \cdot y_i \right)_M = \left(x_{N-1} y_{N-1} (1 + \epsilon) + \left(\sum_{n=0}^{N-2} x_i \cdot y_i \right)_M \right) (1 + \epsilon)$$

$$= \left(\sum_{n=0}^{N-1} x_i \cdot y_i \right) (1 + N\epsilon).$$

It does not come as a surprise that the relative error scales with the size of the vectors that enter the scalar product. The longer the vectors the more calculations are required and the more errors sum up. However, is such an estimate realistic? Unfortunately, it is: Imagine a complex machine learning calculation. In many of these applications, you distribute work among a big computer and you ensure that all components solve kind of the same type of problem. They all solve chunks of the same neural network layer. Often, the individual $x_i y_i$ terms are of the same size. While the summation over the first few thousand terms might still be fine, we will end up in a situation where the accumulated part is already pretty big. Whenever we add further contributions, those are effectively lost (see our programming guidelines before) and thus the ϵ estimate is not too pessimistic at all! The amplification factor in practice might be $N - \hat{N}$ with a certain fixed $\hat{N} \ll N$.

The simple argument on the error bounds already highlight what a more stable implementation looks like. If we don't run through the vector component-by-component but realise the scalar product recursively, we obtain

$$\left(\sum_{n=0}^{N-1} x_n \cdot y_n \right)_M = \left(\sum_{n=0}^{N/2-1} x_n \cdot y_n \right)_M +_M \left(\sum_{n=N/2}^{N-1} x_n \cdot y_n \right)_M.$$

If we assume that the left and the right summand on the right-hand side are of the same order of magnitude, we get an ϵ term due to the addition, but we do not totally lose one contribution. The error term does not increase with N but increases with $\log_2 N$—a significant improvement! We finally bring this together with our previous discussion on vectorised operations:

Blocked scalar product

If your machine has vector units of size B, then it makes sense to split up a scalar product into chunks of size B:

$$\sum_{n=0}^{N-1} x_n \cdot y_n = \sum_{b=0}^{N/B-1} \left(\sum_{n=b \cdot B}^{(b+1) \cdot B-1} x_n \cdot y_n \right).$$

For the inner sum, we use a plain for loop employing vertical vectorisation and even FMA. For the outer sum, we take the block results and add them, but we reduce these partial results along a tree: We sum up the first and second sum over B entries, third and fourth, and so forth. Then, we take the first outcome plus the second one, the third plus the fourth, and so forth.

It is intuitively clear that this blocked version of a scalar product combines the advantages of small relative accumulation errors with high vectorisation efficiency. We preserve the logarithm in the error estimate though this logarithm is now subject to a small factor b.

Some further material to study the topic in-depth

- Robert M. Corless and Leili R. Sevyeri: *The Runge Example for Interpolation and Wilkinson's Examples for Rootfinding*. SIAM Review. 62(1), pp. 231–243 (2020) https://doi.org/10.1137/18M1181985
- Nicholas J. Higham: *Accuracy and Stability of Numerical Algorithms*. 2nd edition. Society for Industrial and Applied Mathematics (2002)
- Lloyd N. Trefethen and David I. Bau: *Numerical Linear Algebra*. Society for Industrial and Applied Mathematics (1997)

> Key points and lessons learned

- Arithmetic stability is a property of the chosen implementation rather than the inherent problem.
- An algorithm's implementation is stable, if its round-off errors remain bounded.
- Stable algorithms give us almost the right answer for the right question. Right question means that we feed the right input data into the algorithm.
- If we encounter unstable operations (equation terms), we can sometimes rephrase them to obtain a higher accuracy.
- Our model problem is locally and globally stable. For iterative or algorithms with many time steps, we however have to distinguish carefully whether we are talking about a single step or multiple steps.

Vectorisation of the Model Problem

9

Abstract

With an understanding how computers handle floating-point calculations, we vectorise our model problem. For this, we first categorise different parallelisation programming techniques, before we commit to a pragma-based approach, i.e. rely on auto-vectorisation plus vectorisation hints to the compiler. Compilers can give us feedback where they succeed to vectorise, which helps us when we introduce OpenMP pragmas to our code. We wrap up the vectorisation topic with a discussion of showstoppers to vectorisation, such as aliasing, reductions, or function calls.

We have discussed so far how a computer handles floating point numbers and how we can tune floating point operations. It is time to translate this knowledge into working code. Before we start to do so, let's brainstorm: How would we like to write fast scientific codes if it were up to us to decide? Would we like to have a totally new programming language fit for the job, or could the idea behind SIMD fit somehow to C? How would we realise vectorisation and what is the right metric to assess whether we have been successful? Will sole runtime measurements be sufficient, or do we need more information to assess the quality of our implementation?

Vectorisation is one flavour of parallel programming. Whenever we write parallel code, there are, in my opinion, four fundamental ways how to do so.

T. Weinzierl, *Principles of Parallel Scientific Computing*, Undergraduate Topics in Computer Science, https://doi.org/10.1007/978-3-030-76194-3_9

Definition 9.1 (*Parallel programming paradigms*)

1. We can rely on the compiler to do the job for us (compiler-based parallelisation). In the vectorisation context, this is also called *auto-vectorisation* or *implicit
   vectorisation*.
2. We can annotate the code to give the compiler hints or recipes how to parallelise.
   In this case, the original/serial code remains unchanged (cmp. compiler-based
   parallelisation) but we manually inject domain knowledge. We augment the code.
3. We explicitly call functions of a parallelisation API, i.e. a specialised library
   behind the scenes.
4. We rewrite everything in a new (parallel) language.

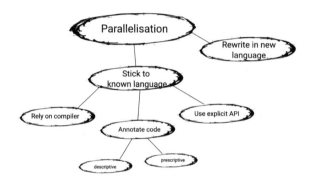

Relying on the compiler to do the work for us seems to be lazy, but very often it
means that we have to revisit our code over and over again to analyse whether the way
we have written down an algorithm allows a compiler to translate it into a parallel
piece of code. Even though we do not directly "add parallelism" to our code base,
the parallelisation does not come for free. It manifests in work to validate if, why and
how a tool manages to create parallel code. On the pro side, as soon as the compiler
yields parallelised code we may assume that it has fewer bugs compared to a manual
parallelisation. We are also platform-independent.

Taking our existing code and adding hints to it is an alternative. We still rely
on the compiler or a tool to do the techi part of producing parallel code, but we
give the tool the domain-specific information explicitly such that it knows what to
do. This can be read as a compromise: We do not outsource all of the work to a
tool which eventually might miss out parallelisation potential, but we also do not
write parallel code ourselves. This means that we have lower development cost
compared to a rewrite, and we reduce the risk to introduce errors—though obviously
our instructions can make the tool do the wrong things. We also avoid the need to
have multiple source code branches if we want to keep the original, serial version.

Another option is a modification of our serial code using a library. We explicitly
call a library to make particular parts of our code run in parallel or to wrap our code
into parallelisation constructs. This is an explicit way how to express parallelism. We
take full ownership. It is also a radical way and requires significant programming.
While it might yield the best performance, it is a strategy that very quickly leads into

situations where our code is tailored towards one parallel computer, and we commit to one library. If the library works only on some machines or its development is discontinued, we have to rewrite.

Finally, we could completely rewrite our code in a new language such as CUDA. This is a strategy we ignore here.

For vectorisation, all three "do not rewrite" variants are on the table if we start from C code. Intel for example provides a set of function calls that we can invoke. They are realised through some assembler code which triggers SSE/AVX instructions. These functions are called *intrinsics*. We will not go down this route, as there are hundreds of these intrinsic functions, they might change from chip generation to generation, and code with intrinsics is not (always) portable to other chips by ARM, AMD or IBM, e.g.

Luckily, compilers today vectorise themselves if we hand them over the right code. We have discussed recipes how to write vector code in Sect. 7.2. As long as we follow these guidelines, we can assume that the compiler does the job for us. Taking into account that vectorisation is not a one-way street performance-wisely, our first objective is to find out what the compiler actually does. After that, we discuss ways how to explicitly tell a compiler to vectorise particular code parts in a certain way. That is, we discuss how we can provide additional information to the translator that helps or forces it to vectorise. For this, we rely on the OpenMP standard. OpenMP (Open Multi-Processing) was originally designed for multicore processors, but since its 4th generation it also supports SIMD. As it is platform-independent, this is our method of choice. Even better, OpenMP's 5th generation (actually already 4.x) support GPGPUs. We have pointed out that major ideas behind GPUs are similar to SIMD. OpenMP will later be able to yield portable vector codes that even fit to GPUs.

9.1 Compiler Feedback

To find out whether vectorisation has been used, we ask the compiler to provide us with feedback. For Intel's compiler, translate your code with

```
icc -qopt-report ...¹
```

You obtain an optimisation feedback file with the extension .optrpt per source code file or some reports on the terminal. It is a plain text file which tells you what the compiler has done. The GNU tools have a similar option (-fopt-info), while

---

[1] In recent versions, Intel have replaced their icc compiler with a new compiler product built upon Clang/LLVM. These compilers are invoked via icx instead of icc. But Intel's feedback is still obtained via -qopt-report.

Clang/LLVM (on Macs) wants to see `-Rpass-analysis=loop-vectorize`
`-Rpass=loop-vectorize -Rpass-missed=loop-vectorize`; all tool-
s have a more fine-granular way to control the output—you have to study the documentation.[2] We obtain feedback like

```
LOOP BEGIN at solution-step1.c(187,3) inlined into myfile.c(266,5)
[...]
remark #15542: loop was not vectorized: inner loop was already \
vectorized
```

which is an example of a loop which hasn't been vectorised as it does not meet
criterion 3 from our list. This is the Intel output of one particular Intel tool generation.
Different tools give you different feedback, but the information published is not that
dissimilar.

Vectorisation steps

Here is a first guideline what we should do from a vectorisation point of view:

1. Any proper optimisation should be driven by profiling. First, we have to find out
   where the majority of our time is spent. Through a profile (it might be reasonable
   to profile for different input data), we get an idea where the CPUhrs are burnt.
2. Next, we use the optimisation report to identify hotspots that suffer from a lack
   of vectorisation. With luck, the report tells us already some of the obstacles (such
   as an embedded if that can not be masked).
3. We tune the code.
4. Before we run our tuned, i.e. optimised code, we recompile with the vectorisation
   report once again to validate that the vectorisation has been successful. This step
   also provides us with estimates how much faster the code should be.
5. Finally, me measure the optimisation's impact on the time-to-solution. Then, we
   start all over again.

Using the compiler feedback all the way through is not just a nice gimnick. As
scientific computing specialists, we should have a good understanding what calculations are done where and which ones are vectorised. This distinguishes us from
trial-and-error coding monkeys.

---

[2] I found on some systems that both Clang and LLVM tend to ignore the analysis request if you
launch it without `-O3` (or another explicit optimisation instruction). So specify an optimisation
level explicitly to be on the safe side.

If the vectorisation report shows us that some parts of the code have not been vectorised though we know it would benefit from vectorisation, it is time to enter phase two of our vectorisation endeavour:

9.2 Explicit Vectorisation with OpenMP

Our explicit, annotation-based vectorisation relies on C pragmas. A pragma is a comment in a special format which passes hints to the compiler how to handle certain code parts. Pragmas are pragmatic, i.e. they allow us to add information to the code that we cannot supplement with built-in capabilities of the language. We augment the code. However, the compiler is free to ignore any pragma if it is not known to the tool. This way, you can insert OpenMP pragmas into your code, and the code still will pass through compilers that do not support OpenMP. There are two different types of annotations:

Definition 9.2 (*Descriptive versus prescriptive annotations*) A *descriptive annotation* gives the compiler hints how to create efficient code. The compiler however is free to ignore these hints. The alternative are prescriptive annotations. A *prescriptive annotation* forces the compiler to parallelise or vectorise, respectively.

OpenMP defines prescriptive annotations, i.e. each OpenMP statement enforces a specific transformation. With a descriptive behaviour, we could rely on the compiler's heuristics and analysis when we vectorise; or parallelise later on. If its internal heuristics suggest for example that a loop is just too small or the thread divergence is too big, then it might ignore the annotation in a descriptive world. We work prescriptive here however. That is, we have to be careful not to break the code and we have to be sure that we actually improve the performance when we insert OpenMP statements. There is no built-in safety net.

OpenMP syntax

Pragmas in C always start with the keyword #pragma. After that, we write which pragma slang we employ. This is omp for OpenMP. Every pragma in this book starts with

```
#pragma omp
```

After the omp identifier, we finally say what our pragma's message is. This is the key instruction. It is called a *directive*. It tells the compiler *what* to do and typically consists of one or few words. Finally, most OpenMP instructions support further parameters that specify *how* a feature is to be realised by OpenMP. These additional parameters are optional. OpenMP calls them *clauses*:

$$\underbrace{\texttt{\#pragma}}_{\text{C's pragma keyword.}} \quad \underbrace{\texttt{omp}}_{\text{OpenMP identifier.}} \quad \underbrace{\texttt{directive}}_{\text{What do to.}} \quad \underbrace{\texttt{clause}}_{\text{How do to it (optional).}}$$

9.2.1 The `simd` Pragma

The SIMD pragma in OpenMP is a prescriptive annotation that always is the preamble to a loop. The loop has to be in a *canonical form*.

```
#pragma omp simd
for (int i=0; i<N; i++) {
  ...
}
```

The example urges the compiler to vectorise this loop. Often, you will get away with these three words. There are however a few further clauses that you can hand over to OpenMP to tweak the vectorisation:

- SIMDsation can be read as loop unrolling. To allow the compiler to do the unrolling "trick", it has to know how often it may unroll the loop. By adding `safelen(length)`, you give the compiler a hint which unroll factor is absolutely safe to use. With this argument, you might get away with a loop range which is not 100% in canonical form. As you manually specify safe loop unrolling counts, the compiler does not have to try to understand some nuances of your code.
- A vectoriser has to assume that each individual vector line has to calculate its own local variables. You might know better, i.e. you might know that a variable within a loop always holds the same value. In this case, tell the compiler about such a local variable via `uniform(myVar)`. Values then are computed only once.
- Vectorisation is particularly fast if you run through multiple arrays and all these arrays are traversed continuously, i.e. one by one in line with the loop counter. Whenever you know that your code runs through arrays in the same way, then you can tell the compiler about this via `linear(myVar)`. It means that you run over the array `myVar` in exactly the same way as you run over the loop index within your for-loop.

- You can add an `if` clause to a `simd` instruction. The clause accepts a boolean expression. If expression evaluates to true at runtime, the loop is vectorised. Otherwise it is not. With `if`, the compilers produces two loop variants: A scalar one and a vectorised one.

Workload and workload character

Search for hints in the vectorisation feedback that tells you whether the compiler thinks of a loop to be too cheap for vectorisation. In this case, you can create one version for the small cardinality and one for reasonably big inputs via the if clause. You might also have expert knowledge about typical vector lengths which helps the compiler.

9.2.2 Loop Collapsing

Nested loops are a natural enemy to vectorisation. Whenever you loop over j and then immediately over i and whenever the loop range of i is independent of j's range, you can permute the loops. Unfortunately, you might not know always which permutation is the better one. You always run risk that the iteration space of the inner loop—and that's the one vectorisation will tackle—is too small to yield efficient vectorisation ranges, yields unfortunate memory accesses, or does something else that is not clever. If you write

```
#pragma omp simd collapse(2)
```

then you tell OpenMP to take the next two for-loops and to collapse them into one big loop. For this, they have to be independent of each other. The pragma yields one large, one-dimensional loop range and thus gives OpenMP more freedom to vectorise.

Nested loops

Search for hints in the vectorisation feedback that tells you that a loop has not been vectorised as it has not been an inner loop. Often, you might have one big loop that has some vector operations inside. Many compilers work greedily: They vectorise the innermost loops and then tell you "sorry, have already vectorised a nested one". It does the tiny little pieces but misses the big picture. Once you manually collapse the loops, vectorisation can make a bigger difference.

Rectangular iteration spaces

When we compute the forces between N particles and if these forces follow $f(x, y) = -f(y, x)$, then it is usually not clever to compute all the $N(N - 1)$ forces. Instead, we compute only half of them and exploit the knowledge that the other half just has the sign inverted. If we visualise this as a matrix, then we are interested in all entries within this matrix besides the ones on the diagonal. As a particle excerts no force on itself, these are by definition zero.

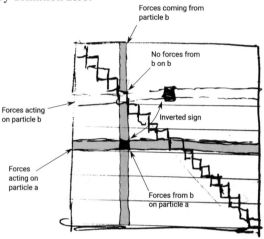

Once we do the $f(x, y) = -f(y, x)$-"trick", we effectively compute only the upper half of this matrix and then wrap it over with a flipped sign. We exploit its anti-symmetry. With SIMD, it is not clear anymore whether this triangularisation is clever. If the particle count is reasonably small, we might be better off with removing our trick. We go back to our rectangular iteration space, i.e. compute both the upper and lower part of the force matrix, but now can maybe collapse the loops over the particles.

9.2.3 Aliasing

A loop body that benefits from vectorisation has to consist of instructions which are independent within the control flow graph: Two instructions can be vectorised, if they are of the same type and if one doesn't feed into the other. We can run instruction A and B in parallel if A's result is not required by B and vice versa.

For a sequence over operations on simple (scalar) variables, it is straightforward for a compiler to extract the control flow graph from your code and, hence, to identify operations that could run in parallel. If they are the same ones, we can vectorise. Things become nasty for C arrays. Arrays in C are identified by pointers, i.e. variables that identify a memory location. You can move these pointers around. Actually, the statement x[i] adds i to the pointer x and then works with the result address. If there are two arrays and accesses x[i] and y[j], the compiler is in trouble: It has no (built-in) guarantee that the two pointers do not point to the same memory location. In such a case, a compiler has to be pessimistic. It has to assume that two pointers from different code locations or loop iterations point to the same address and thus cannot run in parallel.

Definition 9.3 (*Aliasing*) *Aliasing* names the situation that two memory locations can be accessed via different variables (pointers).

Aliasing in a queue

Alisasing is by no means an esoteric problem. Assume you have an array with n elements which we call a queue. Let an operation remove the first element and then move the remaining ones:

```
for (int n=0; n<N-1; n++) {
  x[n] = x[n+1];
}
```

If you instead rely on the more general implementation

```
void foo(int* x, int* y, int newN) {
  for (int n=0; n<newN; n++) {
    x[n] = y[n];
  }
}
```

to move data around, you encounter aliasing. In the former case, the aliasing is apparent. In the latter case, we have to assume it can happen (Fig. 9.1). If we knew however that bar is never invoked on two overlapping memory regions x through x+N and y through y+N, we could vectorise the memory movements.

By definition, a compiler must not vectorise code where aliasing might occur. Otherwise, it might break the code's semantics. If you enable your compiler feedback, it should tell you where aliasing prevents it from generating fast code. Often, these will be false positives. You might know better.

Fig. 9.1 Two arrays of size N=8 are spanned by the pointers x and y. Some entries in memory (filled) can now be reached via x and y (aliasing). A compiler always has to assume that this might happen. It thus cannot reorder or parallelise accesses through x and y

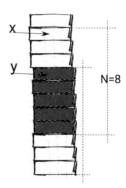

In C, programmers have to tell the compiler explicitly when they are sure that no aliasing happens. There are multiple ways to do so. First, individual compilers like the one from Intel of the GNU suite typically offer special pragmas or annotations to pass on this information. This is a tool-specific solution. Second, almost all C compilers support the `restrict` keyword, even though this keyword is not part of most C standards. To inform a compiler that no aliasing can happen, you can thus add a `restrict` after the pointer.

```
int* restrict myPointer;
```

`restrict` qualifies the pointer, not the data it is pointing to. It means that noone else points into the region accessed through `myPointer`.

Finally, you can use OpenMP's `safelen` qualifier (see remarks when we introduced the `simd` pragma). By notifying OpenMP that it is safe to vectorise over a given chunk within the iteration space, you implicitly inform the translator that aliasing cannot happen.

Be aware of aliasing

Search for hints in the plain vectorisation feedback that tells you that no vector operations have been used as the compiler assumes that arrays (pointer regions) overlap in memory. Work your way backward from here by inserting `restrict` keywords or annotations or `safelen` clauses until the compiler succeeds in vectorising.

There is this myth in the scientific computing community that Fortran is faster than C. When people claim this, they suggest that the C compilers do a poor job benchmarked against their Fortran counterparts. Yet the algorithms are often the same, the implementations strikingly similar, and the target machine is the same as well. One of the reasons that often makes Fortran code indeed faster is aliasing: Fortran is, by construction, avoiding aliasing. As a consequence, the Fortran compiler faces a much easier task when it optimises.

C developers have to manually instruct the compiler about the absence of aliasing. Once they do this, there is no reason why a Fortran compiler's code should outperform a C code. As always, there are a couple of more details and facets to consider when you compare languages and compilers, but aliasing is one of the major players.

9.2.4 Reductions

Reductions, i.e. collective operations computing a result over multiple vector registers, do not allow us to vectorise straightforwardly. With the blocked scalar product plus horizontal SIMD, we however have identified realisation patterns that benefit from vector instructions. It should be possible to gain something from vectorisation even if our code is reduction-heavy. As we do not want to fiddle around with manual rewrites of reductions in code and as reductions are omnipresent in scientific computing, OpenMP provides explicit support for them:

```
#pragma omp simd reduction(+:y)
for (int n=0; n<N; n++) {
  y += x[n];
}
```

At first glance on the control graph, this loop can not benefit from any vectorisation, as each loop iteration adds something to y. The snippet above tells OpenMP explicitly that we reduce the sum over the xs into y via an addition. The compiler now can introduce internal code rewrites—similar to our discussion—and vectorise nevertheless.

The clause's argument supports further basic operations: You can also use the minus sign, the multiplication and logical and bit-wise and (& and &&), or (| and ||) and xor (^). You can also reduce several parameters in one rush; just enlist them separated by a comma. Finally, there is support for min and max.

9.2.5 Functions

Function calls within a loop prevent a compiler to vectorise this loop. A function call has to backup local variables from registers on the call stack, has to create new local variables on the stack, has to store a jump-back address, and so forth. If you however inline a function, then you can get rid of all of this overhead and, consequently, vectorise over functions, too.

Manual inlining is not very attractive for developers. It does not make your code nicer. However, you can inform OpenMP that you intend to use a function within a loop that you want to vectorise later on. The function then is prepared by the compiler. We have to *declare* a function as SIMD-ready:

```
// take element-wise min of entries from a/b and store in y
void min(int* x, int* y, int* z, int N) {
  #pragma omp simd
  for (int n=0; n<N; n++)
    z[n] = min(x[n], y[n]);
}

// properly prepared min function
#pragma omp declare simd
int min (int x, int y) {
  return x < y ? x : y;
}
```

Again, you can augment this declaration with a `simdlen` and thus hand over information on which array areas the routine can work safely. `uniform` is supported, too. The delaration affects only the subsequent function, so you might have to insert multiple declarations into a file: one per function.

In principle, the OpenMP standard does not really care how the compiler reacts to a `simd` declaration, i.e. how the declaration is translated into code. For many mainstream compilers, you will however find that the compiler generates multiple versions of the function for different vector widths. It creates a scalar version (that is the one corresponding to a 1:1 translation), it creates one accepting a vector of two entries, of four, and so forth. When it encouters a function call to a declared function, it thus knows that vector variants of this function are available. Depending on the context, it can select the appropriate one.

While these details are all hidden from the developer—you only declare the function—you have to ensure that a SIMDised function is reasonably simple such that the compiler manages in creating vector variants. Vectorising too complex functions will fail.

Vectorise over standard functions

Many standard functions in C are not SIMD-ready. That is, if you use them within a `simd` loop, the vectorisation has to fail. Examples for this are trigonometric functions such as `sin`. However, vendors often offer SIMD flavours of core (math) routines. You will have to doublecheck whether such variants are available.

Some important material

- OPENMP API Specification: Version 5.0 (2018) https://www.openmp.org/spec-html/5.0/openmpsu42.html
- T. Zeiser, G. Hager, G. Wellein: *Vector Computers in a World of Commodity Clusters, Massively Parallel Systems and Many-Core Many-Threaded CPUs: Recent Experience Based on an Advanced Lattice Boltzmann Flow Solver.* In: W. E. Nagel et al.: High Performance Computing in Science and Engineering '08, Springer, pp. 333–347 (2008) https://doi.org/10.1007/978-3-540-88303-6_24
- G. Hager, G. Wellein: *Introduction to High Performance Computing for Scientists and Engineers.* CRC Press (2010)

> **Key points and lessons learned**

- We have identified three strategies to parallelise C code: auto-vectorisation (implicit), explicit with annotations and explicit through an API.
- There are prescriptive and descriptive annotation techniques. OpenMP is prescriptive.
- Compiler feedback can guide you to pick the prescriptive annotations.
- We have introduced the pragma `omp simd`.
- Restrict annotations, collapsing, reductions and declarations of functions are the four main tools we use when we vectorise.

Wrap-up

The chapters of this Part III orbit around number crunching, i.e. the handling of vast amounts of floating point calculations. We highlight two facets of such calculations: From a numerical point of view, we discuss representation issues—how accurate can we run calculations on our machines—while the speed of floating point calculations is the other aspect of our studies.

We observe that modern machines are well-prepared to handle large calculations efficiently as long as we keep two things in mind: Don't mess up floating point calculations early throughout complex calculations—avoid cancellation or summing up quantities of different magnitude, e.g.—and try to phrase calculations as sequences of small vector calculations while you avoid branching. The two ingredients guarantee that errors do not propagate through your calculations and do not blow up. They also guarantee that the calculations are efficiently piped through the SIMD facilities.

While vector features (and the GPUs' SIMT model) do give us quite a performance boost, exploiting these units is only one way to make a code faster. Vendors today increase the number of cores per node or GPGPU card, too. Our next step hence is to study how to program multicore machines. At the same time, we have so far studied all floating point issues with the assumption that all input data were exact. This is a strong assumption: Input data are also represented by floating point numbers, stem from measurements, or are man-made—very unlikely that they are exact. Besides first multicore upscaling, we thus have to study the effect of input data variations.

There are two important tools that we need to master to achieve our next goals: On the one hand, we estimate how much we actually could gain from a parallel machine. It is one thing to acquire the skills to make a code exploit parallelism. It is another thing to know a priori what might be achievable through which technique—and eventually which technique is the right one to use in which case. On the other hand, we return to our simulation code blueprint and start to argue where the formulae in there come from and whether we can find a (proper) solution at all. This will help us to understand how much of the error that we get is inherent to the problem, and what errors are machine-made and maybe avoidable. Parallelisation and solution plus algorithm properties are totally different things. We therefore treat them in separate manuscript parts from hereon.

Part IV

Basic Numerical Techniques and Terms

Conditioning and Well-Posedness

10

Abstract

Hadamard's definition of well-posedness makes us start to think if a particular problem is solvable on a computer at all. We introduce the terms well-conditioned and ill-conditioned, and finally define the condition number to quantify how sensitive an algorithm is to slight changes of its input. After a brief excursus what the condition number means for linear equation systems—it turns out to be a special case of our more generic definition—we discuss the interplay of the conditioning with forward stability and introduce backward stability as evaluation condition number.

Take a pen into your hand. Your challenge is to bring this pen into a stationary vertical position, i.e. it should not move. Also, you should hold it on one end only.

© The Author(s), under exclusive license to Springer Nature Switzerland AG 2021   119
T. Weinzierl, *Principles of Parallel Scientific Computing*, Undergraduate Topics
in Computer Science, https://doi.org/10.1007/978-3-030-76194-3_10

> There are two solutions to this problem: We can hold the pen at the top and let it dangle. Or we can balance it on our palm to keep it upright. These are the two valid, stationary configurations for the pen. The dangling thing is easy, but we know that balancing it on our palm is tricky. We permanently have to move our hand to keep the pen in position. Once we stop these correcting movements, the pen will fall.

Assume that you put a thermometer into your oven when you bake pizza, and the thermometer shows 220 °C when you switch it off. There is an equation that describes how the oven gradually cools down until it has reached the temperature of your kitchen. Once this equation is written down, we can solve it, i.e. there is a unique solution. The solution even will be relatively insensitive: If our thermometer gets the temperature slightly wrong—for example 222 °C instead of 220 °C—the temperature decrease will not change dramatically. It will not be massively different after an hour from the "right" setup with 220 °C.

Assume you do not know the initial temperature. You however know that the temperature after an hour equals the room temperature. Is it possible to say what the original temperature had been? The answer is no. There are many potential solutions. The oven could have been at room temperature (as we forgot to switch it on). It could have been 220 °C and then we switched it off. Or it continued to heat at 180 °C for a while, we then switched it off, switched it quickly on again at 240 °C and then eventually off. There are infinitely many potential solutions.

It is important to formalise the difference between the two challenges. The first one can be solved by a computer. The second one is tricky, as any solution dumped by a computer could be right or could be totally wrong.

Definition 10.1 (*Hadamard's well-posedness*) A problem is *correctly set*, i.e. well-posed, if

1. it has a solution,
2. this solution is unique, and
3. the solution depends continuously on the input data.

If one of these properties is violated, a problem is *ill-posed*.

Well-posedness is a property that is inherent to the problem setup—we have not yet started to speak about an algorithm. However, we know from the previous sessions that computers make tiny errors all the time. Even worse, we may assume that even our input data are slightly wrong all the time. Consequently, the well-posedness of a problem is extremely relevant for the development of computer codes. We are doomed if we try to make our computer solve an ill-posed problem.[1] But even once we know that a problem is properly set (aka well-posed), we still have to understand how sensitive it is to variations of the input.

Orbit of a satellite

A stable orbit, i.e. an orbit where the satellite neither flies into the void nor crashes into the planet, is very sensitive to the input data: There is a solution, and this solution is unique given the satellite's original position.

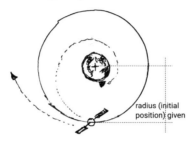

However, once we fix its initial position, only one specific initial velocity ensures that the satellite remains on the stationary orbit. If we slightly alter this one, it follows a completely different trajectory which is nowhere close to the original one. Let the position of a satellite be known for time $t = 0$. Let's furthermore assume that we know the exact velocity that we have to choose to keep the satellite on a stationary orbit, i.e. on a trajectory around the planet with constant radius. A slight reduction of the radius (let's slightly push the satellite towards the planet) or a decrease of the velocity sends the rotating satellite onto a spiral trajectory. It eventually crashes into the planet. If we stay on the stationary orbit but slightly increase the satellite's rotation speed, it will fly away.

An analogous argument holds for the initial position.

10.1 Condition Number

To quantify the sensitivity of a function $f(x)$ to (small) changes of its input, we wobble around with the input and study what happens to the function outcome for a $x + \delta$. We study what impact this small alteration of the input has.

Definition 10.2 (*Ill-conditioned*) If a slight modification of x yields a significantly different $f(x)$, the function $f(x)$ is *ill-conditioned*.

---

[1] As always, there are people developing solutions to ill-posed problems. But that is out of our comfort zone here.

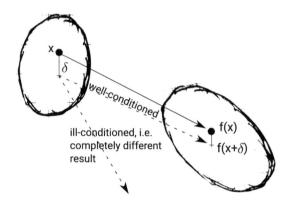

This is an imprecise, non-quantitative statement. Indeed, there's no strict, formal and universally accepted definition of ill-conditioned. Yet, we can introduce a *condition number* which quantifies the behaviour of interest. We can then use a number to argue about a phenomenon quantitatively. Let

$$y = f(x). \quad \text{We now study input data modifications}$$
$$x \mapsto x + \delta \Rightarrow f(x) \mapsto f(x + \delta).$$
$$f(x + \delta) = f(x) + \frac{\delta}{1!}f'(x) + \frac{\delta^2}{2!}f''(x) + \frac{\delta^3}{3!}f'''(x) + \cdots$$
$$\approx f(x) + \delta \cdot f'(x).$$

The formalism ignores higher order terms. It expresses the impact of a change in x on the outcome of f through a first-order Taylor expansion.[2] This allows us to compute the ratio of the relative change in the image to the relative wobbling:

$$cond_x f = \left| \frac{\text{Relative change of output}}{\text{Relative change of input}} \right|$$
$$= \left| \frac{\frac{f(x+\delta)-f(x)}{f(x)}}{\frac{\delta}{x}} \right| \approx \left| \frac{(f(x) + \delta \cdot f'(x) - f(x)) \cdot x}{f(x) \cdot \delta} \right| = \left| \frac{x \cdot f'(x)}{f(x)} \right|.$$

We speak about small changes δ above. To be precise what small means, we normalise it. For $x = 1$, $\delta = 0.1$ is quite significant. For $x = 10^5$, $\delta = 0.1$ is tiny. Dividing by x makes us recognise this fact, though we have to be careful for $x \approx 0$.

The actual function change due to this modification of x enters the enumerator. Again, we are interested in the relative change, i.e. we compare $f(x + \delta) - f(x)$ to $f(x)$. Overall, we put the relative change of the function outcome in relation to the relative change of the input data. As we speak, once more, about small modifications to x, we finally approximate the change in $f(x)$ by Taylor expansion.

---

[2] We revise Taylor expansion in Chap. 11. For the time being, it is sufficient to accept that $f(x+\delta) \approx f(x) + \delta \cdot f'(x)$ is a reasonable assumption here since δ is small.

Definition 10.3 (*Condition number (Definition #1)*) The *condition number* of a function f is

$$\kappa = \left| \frac{x \cdot f'(x)}{f(x)} \right|.$$

κ is traditionally the symbol used for the condition number, and the derivative in the formula above is the derivative w.r.t. the quantity of interest, i.e. $f'(x) = d_x f(x)$. If f depends on many different input variables, we have to study the conditioning for all of these variables one by one, before we make a worst-case statement on the overall condition number. That is, problems can be ill-conditioned w.r.t. particular variables only.

Choice of norms

If we write down the condition number's definition over a scalar function $f(x)$ accepting a scalar input x, we use an absolute value. If we work with vector inputs or vector outputs, we need appropriate vector norms. Therefore, there might be different notions of condition (condition numbers) for different vector norms. We usually use the standard Eukledian norm.

The function f can be complicated. An example is our N-body simulation where f computes the forces from the particles' positions. Due to the normalisation, $x \approx 0$ furthermore is problematic in our ansatz, while $f(x) \approx 0$ is problematic in the final formula. For problematic cases, study the limit and look out for L'Hôpital! Finally, we reiterate that all of these terms are analytical expressions, i.e. we do not use the fact that the computer introduces errors through truncation. We solely argue about the function itself. We therefore say *"a problem is well-conditioned"* or ill-conditioned, respectively. The phrasing highlights that the conditioning is a function/problem property. It has nothing to do with how we code an algorithm. We should never speak of an algorithm or implementation that is ill-conditioned.

There is an alternative definition of well-conditioned, which implicitly defines the condition number: Let's assume that the "wobbling" of the input data is bounded by a certain ϵ. If you are mainly interested in imprecisions as input data are stored in IEEE format, then you can tie it to the machine precision. Otherwise, it is just (yet another) bound. We then quantify the worst-case impact of this error:

Definition 10.4 (*Condition number (Definition #2)*) A function f is *well-conditioned* if there exists a (small) κ such that

$$\frac{|\delta|}{|x|} \leq \epsilon \Rightarrow \frac{|f(x+\delta) - f(x)|}{|f(x)|} = \frac{|f(\delta)|}{|f(x)|} \leq \kappa \cdot \epsilon.$$

The smallest κ is called *condition number*.

The implication in the definition above illustrates that a small relative perturbation δ of the input data which is bounded by ϵ does not arbitrarily mess up the output data.

> **Rule of thumb**

 If we have $\kappa \approx 10^k$, we might loose up to k valid (decimal) digits in our result due to small errors in the input.

The rule emphasises that we always work with "wrong" data in our simulations, as we work on a computer with finite precisions. Therefore, also our input datum has only a certain number of valid bits. In the best case, this number is given by the IEEE standard, but there might be way fewer bits if you work with data from measurements, e.g. κ tells us how many of these valid bits are lost due to our calculations. This statement still might be too optimistic if we pick an unstable algorithm.

While the condition number quantifies how inaccurate our result intrinsically might become—intrinsically means no matter how clever our algorithm and implementation is—a problem with a big κ still might be well-posed according to Hadamard. It is only for $\kappa \mapsto \infty$ that we enter a regime where a small change in the input data triggers a discontinuous change in the outcome. In this case, we violate the third criterion of Hadamard's definition. Overall, we observe that there are well-posed and ill-posed problems. Within the well-posed problems, there are well-conditioned and ill-conditioned ones. The transition is blurry. On the well-conditioned side, there are algorithms which are stable—so far we know only stability w.r.t. the round-off error though we'll introduce a second notion of stability soon—and then there are algorithms/implementations which are not stable. Among the stable algorithms, there are efficient and not that efficient implementations (Fig. 10.1).

With all of this in mind, we can sketch what a programmer has to take into account when she starts to write a simulation code:

Fig. 10.1 Revised classification of problem, algorithm and implementation properties including the notion of well-conditioned (cmp. Fig. 8.2 which did lack these details yet highlights that some distinctions are blurry)

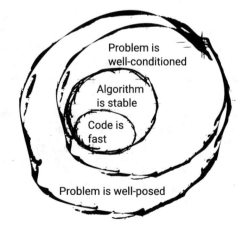

1. Check whether a problem is well-posed. If it is ill-posed, we basically can stop right away. We have to study another problem.
2. Check whether a problem is well-conditioned. The bigger the condition number the more carefully we have to be, when we make statements about the algorithms' outcome.
3. Check whether our implementation is stable from a round-off point of view.
4. Check whether we deliver an efficient implementation.

All of our statements can be input-dependent: A problem might be well-posed and well-conditioned for a given set of input parameters, while it breaks down for others. The same holds for the stability. A classic case for the former is a condition number that depends on the number of inputs: A simulation could be well-conditioned for small number of input data (particles) but its condition number could grow quickly with a growing input size.

10.2   The Condition Number for Linear Equation Systems

Whenever you work with linear equation systems, a lot of literature relies on a specialisation of the condition number. While not adding anything substantially new, it is good to have this specialisation in mind (Fig. 10.2), as linear equation systems are omnipresent.

With linear equation systems, your $y = f(x)$ solves $Ay = x$. You are interested in the value $y = A^{-1}x$. The statement assumes that A can be inverted. This allows us to characterise the κ: A (square) matrix A takes a vector and maps it onto another vector. This mapping can be a rotation, stretching, mirroring, and so forth. If a vector's direction remains preserved, we call it an eigenvector v with an eigenvalue λ:

$$Av = \lambda v.$$

Given a diagonalisable matrix A, we can decompose any input vector x into a linear combination of the eigenvectors: $x = a_1 v_1 + a_2 v_2 + \cdots$ The v_i are the eigenvectors,

Fig. 10.2 Any vector in the 2d plane can be represented by a linear combination of two vectors v_1 and v_2. Let Ax squeeze v_1 and leave v_2 intact. If you pick a b in $Ax = b$ along v_2, any slight alteration along v_1 yields a massive difference in the preimage

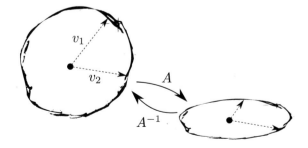

and the a_i are scalar weights. This yields a decomposition into vector components for which we know which ones are particularly amplified: $Ax = A(a_1v_1+a_2v_2+\cdots) = a_1\lambda_1v_1+a_2\lambda_2v_2+\cdots$ The weights a_i which correspond to large eigenvalues v_i make a significant contribution to the outcome. The weights corresponding to very small eigenvalues do not.

Singular value decomposition

Our arguments rely on the fact that we have n positive eigenvalues

$$0 < \lambda_{\min} \leq \cdots \leq \lambda_{\max},$$

with eigenvalues $v_{\min}, \ldots, v_{\max}$. Whenever this is not the case, we have to use the singular values.

We are not really interested in Ax, but we exploit that $\lambda_i(A^{-1}) = \frac{1}{\lambda_i(A)}$. If we know the eigenvalues of A, we know the eigenvalues of A^{-1}, too. Returning to our condition number discussion where we want to solve $y = A^{-1}x$, the worst-case becomes apparent: We could pick an input x which is very close to the biggest eigenvector from A^{-1}. Then, we wobble around x by a delta δ which is close to the smallest eigenvector.

We pick our ϵ such that $|\delta/x| = |v_{\min}(A^{-1})|/|v_{\max}(A^{-1})| \leq \epsilon$. For this particular input, let's quantify the calibrated output $|f(\delta)|/|f(x)|$ using the eigenvalues from A:

$$\frac{|f(\delta)|}{|f(x)|} = \frac{|A^{-1}\delta|}{|A^{-1}x|} = \frac{|\lambda_{\min}(A^{-1})\delta|}{|\lambda_{\max}(A^{-1})x|} = \frac{|\lambda_{\max}(A)\delta|}{|\lambda_{\min}(A)x|} \quad \text{and therefore} \quad \kappa = \frac{\lambda_{\max}(A)}{\lambda_{\min}(A)}.$$

Inverting A is what our algorithm for $f(x)$ is responsible for. We do not know A^{-1}. However, our theorem requires only statements on the eigenvalues of A. For many problems, there are good a priori estimates of $\lambda_{\max}(A)$ and $\lambda_{\min}(A)$. We can provide an estimate for κ.

10.3 Backward Stability: Arithmetic Stability Revisited

Our introduction of the condition number and stability carefully distinguishes the two things. Stability is about an implementation and conditioning about the problem itself. Condition number and round-off stability however have something to do with each other. When we introduce the term forward stability in Chap. 8, we emphasise that we work with exact input data. It is the inexact arithmetics within the machine that nevertheless pollute the outcome, and we end up with something which is not precisely what we wanted to compute.

This is one way to read the error introduced by the machine. Backward stability is another way, where we interpret numerical errors throughout the calculations as modifications of the input data. It goes back something some books call Wilkinson's principle. We ask ourselves whether we can read the outcome of a computation as the exact solution for different input data: We obtain something slightly off. Yet, this is the exact solution for a slightly different problem!

With this interpretation, the question is whether we have solved a reasonably close alternative problem, or whether we've solved something completely different and unrelated. As long as the round-off errors have made us solve a problem that's close to the problem we originally wanted to solve, our algorithm is (backward) stable. Otherwise, it is not. The name backward stable comes from the fact that we relate the outcome to where we come from, i.e. we look backwards.

Definition 10.5 (*Backward stable*) Let $f_M(x)$ be the outcome of an exact input x computed through f with machine precision. That is, f_M suffers from some round-off error. Determine $\hat{x}$ that would yield f_M in exact arithmetics. An algorithm is *backward stable* iff

$$\exists C : \forall x \text{ with } f_M(x) = f(\hat{x}) : \frac{|\hat{x} - x|}{|x|} \leq C\epsilon.$$

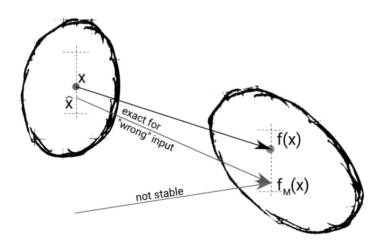

With the backward stability analysis, we search for a magic constant C such that—no matter which input x we select—we can rewrite the outcome as the exact solution for a perturbed $\hat{x}$ and this $\hat{x}$ is not too far away from x. Too far means C times the machine precision.

Square root

Let's compute $f(2) = \sqrt{2} = \sqrt{1.0 \cdot 2^1}|_2$ (with an unbiased exponent). The result is $f(2) = 1.414213\ldots = 1.0110101000001001\ldots|_2$. If we work with half precision and if we assume that our square root implementation (which is typically a function in a programming language) yields the "right" result, we get $y_M = f_M(2) = 1.0110101000|_2$. This is $1.4140625|_{10}$ which gives us $(1.4140625|_{10})^2 = 1.99957275390625_{10}$. Instead of speaking of a (forward) round-off error introduced by squeezing $\sqrt{2}$ into IEEE format, we read this result as "well, we've returned the right square root value but just for a slightly smaller input value than the one we'd been supposed to handle".

With this backward idea, we still break down an algorithm into a sequence of elementary steps, i.e. unary and binary calculations. Instead of accepting that each step introduces some error, we assume that each step computes exactly the right result yet works on some altered data. This data is off by some value.

Backward stability illustrated

We interpret the function evaluation f_M as precise, i.e. not polluted. It has however been given a slightly biased input which is off from the real intermediate result by $\tilde{x} - x$. This offset models the truncation error within the code of f_M as a priori alteration of the input fed into f.

The algorithm is backward stable if this "being off" is not amplified too much—but that's our definition of conditioning: Let an algorithm be a sequence of functions. Each function has a condition number. If a sequence of function evaluations does not yield a condition number that's unbounded, then the initial error is not amplified too much. If only one step is ill-conditioned, then the overall algorithm is ill-conditioned and there is no stable implementation. For this reason, some scientists avoid the term backward stable and instead speak of the *evaluation condition number*: It is not about the condition number of the problem but about the condition number of the individual evaluation steps that we have chosen when we wrote down the algorithm.

Some important material

- Robert M. Corless and Leili R. Sevyeri: *The Runge Example for Interpolation and Wilkinson's Examples for Rootfinding*. SIAM Review. 62(1), pp. 231–243 (2020) https://doi.org/10.1137/18M1181985
- Felix Kwok, Martin J. Gander, Walter Gander: *Scientific Computing—An Introduction Using Maple and MATLAB*. Springer (2014)
- Nick Higham: *What Is a Condition Number?* "What Is" Series. https://nhigham.com/2020/03/19/what-is-a-condition-number (2020)
- Nick Higham: *What Is Backward Error?* "What Is" Series. https://nhigham.com/2020/03/25/what-is-backward-error (2020)

> **Key points and lessons learned**

- Hadamard's definition of well-posed problems.
- The formal definition of the condition number.
- Our discussion clarifies that we search for stable algorithms within well-conditioned problems of well-posed challenges.
- Maximal and minimal eigenvalue of a matrix allow us to write down the condition number for solvers of linear equation systems.
- Besides forward stability there is also backward stability which interprets the flawed outcomes to a problem due to a machine with finite precision as the exact outcome to a slightly different problem.
- There is a rule of thumb how much precision we loose inherently for a given problem with a given condition number. As the condition number is a problem property, we cannot do better than that in our implementation.

Taylor Expansion

11

Abstract

We revise the Taylor series and Taylor expansion. The text's intention is to remind us of this important tool, and to refresh some important technical skills such as Taylor for functions with multiple arguments. We reiterate the smoothness assumptions we implicitly employ. This chapter revisits Taylor expansion in a very hands-on fashion. Feel free to skip the whole chapter if you don't have to refresh your knowledge.

Imagine you are driving with your car along a road through a national park. You know that you are 100 m above sea level. Where will you be in around five minutes' time? Assume the road is completely dark before you. Without further information, the best answer you can give is "guess we will still be at 100 m".

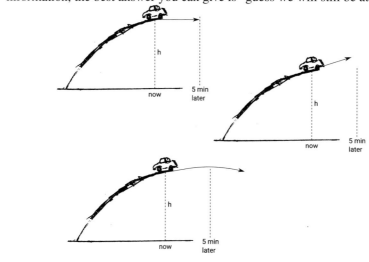

© The Author(s), under exclusive license to Springer Nature Switzerland AG 2021
T. Weinzierl, *Principles of Parallel Scientific Computing*, Undergraduate Topics
in Computer Science, https://doi.org/10.1007/978-3-030-76194-3_11

Let's assume you are at 100 m at the moment, but the road is ascending. You are climbing 5 m per minute. A more sophisticated answer now is "I guess we'll be at 125 m in 5 min from now". You take the knowledge about the current climb into account.

Let's finally assume that we are at 100 m, the road is ascending, but we see that the slope starts to decrease. There's actually a plateau in front of you. Without further information, you might assume that the road continues to descend after the plateau and eventually leads right into the sea. What does a sophisticated prediction now sound like?

Whenever we want to argue about some functions, we first have to make a choice how to write down these functions. One of the simplest ways is to write them down as polynomials. In this case, functions look similar to

$$f(x) = a_1 \cdot x, \qquad \text{or}$$
$$f(x) = a_2 \cdot x^2 + a_0.$$

This type of ansatz is also called *power series*.

Definition 11.1 (*Linear combination over basis*) If we write down a function as

$$f(x) = \sum_i a_i \cdot \phi_i(x),$$

we write the function down as a *linear combination* over *shape functions* $\phi(x)$. The *weights* a_i are independent of x, i.e. fixed, yet obviously depend on $f(x)$ and the choice of $\phi(x)$.

Converting functions into a linear combination of relatively simple shape functions is a powerful technique. We can choose the ϕs such that we know something about their properties. We can for example pick functions where we know the derivative. Once the shape functions are fixed, the a_is adopt. We calibrate them towards our function of interest. From hereon, we derive statements on the total behaviour of the function from the weighted "sum of the properties": If we want to know the overall derivative, e.g., we compute the derivative per shape and then sum these contributions up subject to the a_is.

Definition 11.2 (*Power series*) If we work with polynomials, our shape functions are $\phi_i(x) = x^i$, i.e. they are from the set $\{1, x, x^2, x^3, \dots\}$. The functions ϕ_i form a *basis*, as no two ϕ_is can be combined into a third one. The (infinitely long) linear combination of these shape functions yields a *power series*.

Other popular bases

Another popular basis in scientific computing is $\{\sin(\pi x), \cos(\pi x), \sin(2\pi x), \cos(2\pi x), \sin(4\pi x), \ldots\}$. We know this one as Fourier basis. It assumes that a signal, i.e. the function $f(x)$, is periodic over $(0, 2\pi)$ and can be represented as a linear combination of waves. Another example for a basis is B-splines. They are popular in computer graphics. The clue here is that each spline is defined over a subinterval (t_i, t_{i+1}). Outside of this interval, it is zero. A linear combination of such splines thus is a concatenation of curve segments.

Assume we have points given in a coordinate system. We want to draw a function through these points. They are given as $f(x_0)$ for a point x_0, $f(x_1)$ for a point x_1, and so forth.

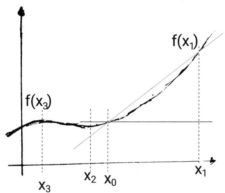

For the Taylor series[1], we assume that our functions look like $f(x) = \sum_{i \geq 0} a_i x^i$. Pick one of our points. Let this one be the x_0 with $f(x_0)$. The best approximation, i.e. the best function that runs through this point is $f(x) = f(x_0)$. We set $a_0 = f(x_0)$, and all the other as zero. This is a constant function, and the best we can do for only one single sample point. We have no further data.

Next, we use $f(x_1)$, too. Unless $f(x_1) = f(x_2)$, which would be a coincidence, our initial fit with $f(x) = a_0 \neq f(x_1)$ for $x = x_1$. So we have to use one more function from our basis:

$$f(x) = a_0 + a_1 x.$$

Once we insert x_1 and the known a_0, we obtain a line which is a good approximation for two given points. We continue with the other points.

The above process is called *interpolation* as we use a set of (simple) basis functions to successively approximate a function. Due to the fact that we know our basis functions, we get a feeling what the solution in-between given sample points looks like. We interpolate.

---

[1] As this text is written at Durham University, it is only fair to emphasise that Taylor's mother was from Durham, i.e. there is Durham heritage in these formulas.

Unless we are lucky and hit multiple sample points "prematurely" (if all points have the same value, picking a_0 right makes us meet all of them in one go), we need as many basis functions as we have points. If we have more points than basis functions—as we stick to a constant number of them, e.g.—we have an *overdetermined* system, i.e. we have to find weights such that $f(x)$ is approximated as good as possible. At the same time, we have to accept that we will not be able to match all given sample points. If we have more shape functions than sample points, then we have an *underdetermined* system. Many a_i combinations then interpolate all sample points. We have to come up with a plan which combinations are reasonable ones.

Machine learning

Interpolation is the backbone of machine learning: Each layer of a neural network yields a large linear combination of simple, elementary (shape) functions. The total function is a linear combination of these elementary ones. The "training" of a machine learning model is finding proper weights a_i.

11.1 Taylor

In my introduction of power series as interpolating functions, we select the weights such that the resulting $f(x)$ runs through a given number of sample points. There is a different approach to determine the weights.

Definition 11.3 (*Taylor series*) The *Taylor series* for a function $f(x)$ is a power series where the a_is stem from the derivatives of $f(x)$.

If we assume that our function is represented as

$$f(x) = \sum_{i \geq 0} a_i x^i,$$

we can directly write down the derivatives of this function[2]:

$$f(x) = \sum_{i \geq 0} a_i x^i,$$

$$\partial_x f(x) = \partial_x^{(1)} f(x) = \underbrace{\partial_x(a_0)}_{=0} + \underbrace{\partial_x(a_1 x)}_{=a_1} + \underbrace{\sum_{i \geq 2} i a_i x^{i-1}}_{2a_2 x + 3a_3 x^2 + 4a_4 x^3 + \dots} = \sum_{i \geq 1} i a_i x^{i-1},$$

$$\partial_x \partial_x f(x) = \partial_x^{(2)} f(x) = \sum_{i \geq 2} i(i-1) a_i x^{i-2}, \text{ and so forth.} \tag{11.1}$$

---

[2] See the remarks on derivative notations in Sect. 3.1.

With this formalism, we can fiddle in weights a_i in the following way: Let's assume that we know a function in the point $x = 0$, i.e. we know $f(0)$. Then, we automatically know $a_0 = f(0)$ in (11.1). Let's assume that we furthermore know something about $\partial_x f(x = 0)$. We do not know what the function overall looks like, but we know its value in $x = 0$ plus its derivative. In this case, we can plug our match $a_1 = \partial_x f(x)$ into the formula above.

Definition 11.4 (*Taylor expansion*) The *Taylor expansion* takes the function $f(x)$ and approximates it with a constant function. It then expands this approximation further and further by adding more basis functions from the power series. The corresponding weights are given by the derivatives of the function $f(x)$ that is approximated. The (infinite) result of the *Taylor expansion* is the *Taylor series* for the function $f(x)$.

Taylor expansion versus power series interpolation

Straightforward interpolation means fitting a_is such that a function runs through a given set of points. With Taylor expansion, we use the same paradigm to match a function's value plus its derivatives in one point.

Taylor takes a function value plus an arbitrary set of derivatives in one point, and it provides us with an idea what the function around this point looks like. If we are given only a point, we assume the environment is represented by a constant function. If we are given with a point plus its derivative, we end up with a linear extrapolation. If we are given a point, its derivative and a second derivative, we obtain a parabola.

The Taylor series for a function around a point x equals

$$f(x + dx) = f(x) + dx \cdot \partial_x f(x) + \frac{dx^2}{2} \cdot \partial_x \partial_x f(x) + \frac{dx^3}{3 \cdot 2} \cdot \partial_x \partial_x \partial_x f(x) + \dots$$

$$= \sum_{i \geq 0} \frac{dx^i}{i!} \partial_x^{(i)} f(x)$$

This is a formula that we will need over and over again throughout this text. Its derivation is a straightforward result of the Taylor expansion, i.e. we repeat the steps from (11.1) centred around x. The point $f(x)$ is fixed, and we study the function in a dx-environment around x. Though we prefer the arrangement above, a reordering highlights the connection to a power series:

$$f(x + dx) = \underbrace{f(x)}_{:=a_0} + \underbrace{\partial_x f(x)}_{:=a_1} \cdot dx + \underbrace{\frac{\partial_x \partial_x f(x)}{1 \cdot 2}}_{:=a_2} \cdot dx^2 + \underbrace{\frac{\partial_x \partial_x \partial_x f(x)}{1 \cdot 2 \cdot 3}}_{:=a_3} \cdot dx^3 + \dots$$

Notation remark

The convention $0! = 1$ for the factorial is it that enables us to write down the Taylor series in its compact form. That's what a good convention looks like!

There are three important assumptions tied to the Taylor expansion and the way we use it:

1. We assume that the function is smooth, i.e. we assume that all the derivatives $\partial_x^{(n)} f(x)$ do exist. No $\partial_x^{(n)} f(x)$ does jump. Otherwise, the $\partial_x^{(n+1)} f(x)$ would not be there.
2. We assume that the sum over all the derivatives remains finite. In particular, we assume that the $f(x)$ does not run into $\pm\infty$. This implies that the a_i very quickly become very small.
3. We assume that the a_i, once fitted, are constant: We fit them around x through the derivatives of $f(x)$ and then keep them invariant. The development of the function around $f(x)$ thus depends solely on the distance.

Since the magnitudes of the a_is decrease, we are on the safe side if we cut off the Taylor series: The "higher i" terms are scaled with dx^i, their weights a_i are fixed, and the $\partial_x^{(i)} f(x)$ are bounded. Due to the dx^i scaling, we can cut off. As we however assume that the a_i are constant once we have fitted them to the derivatives of $f(x)$ at a point x, we should only make statements about $f(x)$ in a small dx-area around x.

11.2 Functions with Multiple Arguments

We have so far used Taylor as a one-directional thing. If we have a function $f(x, y)$, we can use Taylor for either coordinate. That is, we can either develop Taylor for x or for y. It is a little bit like the Manhatten distance or an old 2d computer game where you can either go left/right or up/down:

$$f(x + dx, y + dy) = f(x, y + dy) + dx \cdot \partial_x f(x, y + dy) + \dots \quad \text{or} \quad (11.2)$$
$$f(x + dx, y + dy) = f(x + dx, y) + dy \cdot \partial_y f(x + dx, y + \dots$$

Once we have committed to one direction and have obtained the Taylor expression, we can next run in the other direction. Let's assume we have commited to the upper row first. Once the Taylor over x is written down, we take (11.2) and apply the expansion again for dy. The terms on the right-hand side in the first line of (11.2) expand to

$$f(x, y + dy) = f(x, y) + dy \cdot \partial_y f(x, y) + \ldots$$
$$\partial_x f(x, y + dy) = \partial_x f(x, y) + dy \cdot \partial_x \partial_y f(x, y) + \ldots$$
$$\partial_x^{(2)} f(x, y + dy) = \partial_x^{(2)} f(x, y) + dy \cdot \partial_x^{(2)} \partial_y f(x, y) + \ldots,$$

and we can start to reorder all the partial derivatives in the overall expression. We obtain

$$f(x + dx, y + dy) = \left(f(x, y) + dy \cdot \partial_y f(x, y) + \ldots \right)$$
$$+ dx \cdot \left(\partial_x f(x, y) + dy \cdot \partial_x \partial_y f(x, y) + \ldots \right) + \ldots$$

It is convenient to revisit our notation. Let's write down $f(x, y)$ as a function with only one parameter

$$f \begin{pmatrix} x \\ y \end{pmatrix},$$

which is a vector. Hence, we study

$$f \begin{pmatrix} x + dx \\ y + dy \end{pmatrix} = f \begin{pmatrix} x \\ y + dy \end{pmatrix} + dx \cdot \partial_x f \begin{pmatrix} x \\ y + dy \end{pmatrix} + \ldots$$

$$= f \begin{pmatrix} x \\ y \end{pmatrix} + dy \cdot \partial_y f \begin{pmatrix} x \\ y \end{pmatrix} + \ldots + dx \cdot \partial_x \left(f \begin{pmatrix} x \\ y + dy \end{pmatrix} \right) + \ldots \text{ roll out left term}$$

$$= f \begin{pmatrix} x \\ y \end{pmatrix} + dy \cdot \partial_y f \begin{pmatrix} x \\ y \end{pmatrix} + dx \cdot \partial_x \left(f \begin{pmatrix} x \\ y \end{pmatrix} + dy \cdot \partial_y f \begin{pmatrix} x \\ y \end{pmatrix} + \ldots \right) + \ldots$$

$$= f \begin{pmatrix} x \\ y \end{pmatrix} + dy \cdot \partial_y f \begin{pmatrix} x \\ y \end{pmatrix} + dx \cdot \partial_x f \begin{pmatrix} x \\ y \end{pmatrix} + dx \, dy \cdot \partial_x \partial_y \ldots$$

$$= f \begin{pmatrix} x \\ y \end{pmatrix} + \begin{pmatrix} dx \\ dy \end{pmatrix} \cdot \underbrace{\begin{pmatrix} \partial_x f((x, y)^T) \\ \partial_y f((x, y)^T) \end{pmatrix}}_{=: \nabla f((x, y)^T)} + \ldots$$

$$\underbrace{\phantom{= f \begin{pmatrix} x \\ y \end{pmatrix} + \begin{pmatrix} dx \\ dy \end{pmatrix} \cdot \begin{pmatrix} \partial_x f((x, y)^T) \\ \partial_y f((x, y)^T) \end{pmatrix}}}_{\text{scalar product}}$$

The vector Taylor expansion thus can be written down just like the normal Taylor expansion: The scalar step size becomes a vector. The first derivative is a vector of the partial derivatives called the gradient, and we multiply it with the step vector via a scalar product. The second derivative would then a matrix of second derivatives multiplied with the step vector twice such that it "collapses" into a scalar.

Partial and total derivatives

If one of your arguments in $f(x, y)$ depends on the other, you have to ensure that you work with total derivatives in your Taylor expansion.

11.3 Applications of Taylor Expansion

We intuitively use Taylor for our planet model in Chap. 3, where we assume that we can represent the planet position with a Taylor series. As the velocity per object is stored, it determines (fixes) our a_1. Then, we throw away all the terms a_i with $i \geq 3$, i.e. we set them zero. This leaves us with a_2 into which we plug our force equations. The example illustrates that we only should use small time step sizes, as the velocity per object is not constant but also subject to the forces. Fixing it thus is an approximation.

If we want to find out how a function reacts to tiny variations around a value, i.e. how it reacts to pollution of the input values or computational errors, we can recast the behaviour around the value into a Taylor series. We assume that the $|a_i|$s decrease rapidly and throw all the stuff away that is associated with higher order derivatives. Consequently, we end up with a description of the function's behaviour around a value which is shaped only by the function value plus its first derivative. This is exactly the mindset we have employed for our condition number definition.

Revision Material

- Walter Rudin: *Principles of Mathematical Analysis*. 3rd ed. McGraw-Hill (1976)

> Key points and lessons learned

- We can derive the Taylor series. However, we should memorise as it is such an important, omnipresent tool.
- Basis, shape (or basis) functions and their linear combination subject to weights are important techniques that we use all the time.
- The Taylor expansion yields the Taylor series of a function around a given value, and it uses a power series as starting point.
- We usually chop this series off after a few terms.
- Our analysis of the condition number and our round-off analysis can be explained through a Taylor ansatz.
- Whenever we cut off a Taylor series, i.e. neglect terms, it is important to keep in mind that we rely on rather restrictive properties of the underlying function. Notably, we expect the function to have an arbitrary number of derivatives and we expect that the impact of a derivative on the function nearby is the smaller the higher the derivative.

Ordinary Differential Equations

12

Abstract

There is a lot of definitions, terminology and theory around ordinary differential equations (ODEs). We introduce the terminology and properties that we need for our programming as well as the knowledge that we can exploit when we debug and assess codes. With the skills to rewrite any ODE into a system of first-order ODEs and the Taylor expansion, we finally have the mathematical tools to derive our explicit Euler formally. This chapter revisits some very basic properties of ordinary differential equations (ODEs). Feel free to skip the majority of this chapter and jump straight into Sect. 12.3, where we bring our ODE knowledge and the ideas behind Taylor together.

For those who fancy a revision, we start once more with a pen 'n paper experiment: Let there be a radioactive substance which is quantified by $r(t)$, i.e. it is given as a value r over time. Assume we have 16 nuclei initially. That is $r(t = 0) = 16$. After 1s (I omit the units from hereon), only half of them are left. $r(1) = r(0)/2 = 8$. After another second, again half of them have vanished and we consequently have $r(2) = 4$.

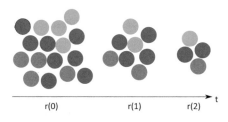

> The pattern here is obvious: Every time unit of 1, we loose half of our substance. This can be described as $r(t) = r(t-1)/2$, but what happens if the t is continuous and the $r(t)$ is continuous as well? What if we are interested in $r(0.5)$?
>
> The continuous counterpart of our recursive equation is
>
> $$\partial_t r(t) = \alpha r(t) \qquad \text{with an } \alpha < 0.$$
>
> The evolution of the quantity $r(t)$ in time, i.e. its time derivative, depends on the quantity itself. If we have twice the amount of the substance, we also loose twice the amount of substance over a given time span.

Chapter 3 introduces equations where the evolution of the equation does not only depend on an explicit argument such as time, but also on the solution itself. Our evolution of the planets, i.e. how they move, depends on both time and the arrangement of the planets themselves.

Definition 12.1 (*Ordinary differential equation*) An equation of the type

$$F(t, f(t), \partial_t f(t), \partial_t \partial_t f(t), \ldots, \partial_t^{(n-1)} f(t)) = \partial_t^{(n)} f(t) \qquad (12.1)$$

with variables $t \in \mathbb{R}$ and functions $f(t)$ is an *ordinary differential equation* (ODE).

We notice that this is a different type of equation compared to what we know from school. Its unknowns are both the traditional parameters (here t) as well as functions and their derivatives. That is, we are not searching for a particular argument here—if we solve $y(x) = x^2 = 16$ we want to obtain $x = \pm 4$—but we are searching for a whole function $f(x)$ that materialises a solution, i.e. fits to the ODE. This function usually depends on t, too.

Our definition contains only time-derivatives. There are also differential equations that have time derivatives plus spatial derivatives or, more general, derivatives along more than one direction. Such equations are *partial differential equations*. They are not subject of study here.

Our planet trajectories are one example for an ODE. The introduction's decay equation is another one; one for which we can even write down a solution. It has solutions $r(t) = r_0 \cdot e^{\alpha t}$ as we quickly validate:

$$r(t) = r_0 \cdot e^{\alpha t} \qquad \text{is our ansatz (educated guess) with}$$
$$\partial_t r(t) = \alpha\, r_0 \cdot e^{-\alpha t}, \text{ i.e.}$$
$$\underbrace{\partial_t^{(1)} r(t)}_{=: \partial_t^{(1)} f(t)} = \underbrace{\alpha \partial_t^{(0)} r(t)}_{=: F(t, f(t))} .$$

What we do here is a classic: We rely on our intuition and come up with an ansatz: "this is what the solution could look like, but there are many free parameters". Next, we validate that this educated guess fits to the ODE. In our example, it does. The $r(t)$ is the $f(t)$ in the definition, while the n from the definition equals $n = 1$. We also see that t does not enter F directly in this particular case. We have $F(t, r(t)) =$

$F(r(t))$. Finally, we take the two free parameters r_0 and α of our ansatz and plug in $r(0) = r_0 = 16$ and $r(1) = 8 = 16e^{\alpha}$.

This is an approach that yields an analytical solution. It yields a write-up of a "proper" function, i.e. something we can simply evaluate. However, constructing analytical solutions is not the strategy we follow in this course. We want a computer to construct approximations for complex ODEs where we struggle or fail to write down analytical solutions.

12.1 Terminology

Even though we are not interested in the symbolic approach to solve ODEs, we need a certain set of terms and concepts to discuss solutions properly. Furthermore, the terms often allow us to validate whether a numerical solution is reasonable—if we know that a solution should have certain properties, we can look out for them in our plots, e.g.

Before we start, we have to clarify whether and where a solution of an ODE does exist. Otherwise, it is meaningless to even boot our computer. We assume that there is a certain area of t and $f(t)$ for which the $F(\ldots)$ is well-defined and a solution does exist. That means that the solution is reasonably smooth, i.e. its derivatives do exist. You can not write down a term $\partial_t f$ if the function exhibits a zigzag pattern (Fig. 12.1).[1]

Within the area where everything is well-defined, we are typically interested in a particular solution:

Definition 12.2 (*Initial value problem*) If we search for a solution of an ODE fitting to a certain $f(0) = f_0$, we speak of an *initial value problem* (IVP).

Fig. 12.1 A function with a zigzag pattern. It is clear that the first derivative is not defined where the curve turns around

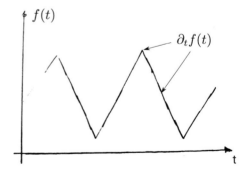

---

[1] There are weaker notions of derivatives such that certain jumps (in the derivatives, e.g.) are fine. But that goes way beyond the maths that we do here.

In this book, all of the problems we study with our code are IVPs. This does not come as a surprise, as our numerical computer programs need a (floating point) value to start from. Next, we discuss the order of an equation:

Definition 12.3 (*Order*) The n in (12.1) is the *order* of the ODE.

Standard equations from school such as $y(x) = x^2$ can be read as degenerated ODEs. We can call them 0th-order ODEs. If a problem has only first derivatives, we speak of a problem given in *first-order formulation*.

The ODE's solution f can be a scalar or a vector function, i.e. it can either return a single value or a whole vector. In this context, there is a nice technique which we can use for all ODEs:

Rewrites of higher-order ODEs

Let's assume that we have a second-order, scalar ODE to solve:

$$F(f(t), \partial_t f(t)) = \partial_t^{(2)} f(t),$$

where $f : \mathbb{R} \mapsto \mathbb{R}$ accepts a scalar and returns a scalar. We introduce a helper function $f_2(t) := \partial_t f(t)$. This is not a free function in the sense of a free parameter in an ansatz. It is a function which always "displays" the derivative of $f(t)$ at a given point in time. With this helper function, we write the whole thing down as a system of two equations:

$$\begin{pmatrix} \partial_t f \\ \partial_t f_2 \end{pmatrix}(t) = \begin{pmatrix} f_2(t) \\ F(f(t), f_2(t)) \end{pmatrix}.$$

The first line is the newly introduced helper function. In the second line, we take our original problem formulation and replace $\partial_t f(t)$ with the helper. To make things more convenient, it might furthermore be reasonable to rename $f(t) = f_1(t)$. We still have to solve a system of ODEs, but this system is first order. Though this is a purely mechanical replacement exercise—and thus not particularly exciting from a math's point of view—it is a nice insight for us as programmers: It is totally sufficient to study algorithms for first-order ODEs! If we write a code that can solve a system of first-order ODEs, then this code can be used to tackle any order.

We use this reformulation already: A planet in space is completely characterised for a given time $t = 0$ by its initial position and velocity. These six quantities form the initial value of the IVP. Once released, a planet's acceleration, i.e. the time derivative of the time derivative of the position, depends on the other planets around. The acceleration depends solely on the geometric arrangement of all objects. The movement equation hence is determined by a second-order ODE where $\partial_t \partial_t f(t)$ depends on $f(t)$. If you have N objects, each one described by a vector in space, then you end up with $3N$ second-order ODEs that all depend upon each other: $f(t) : \mathbb{R} \mapsto \mathbb{R}^{3N}$ maps time onto positions.

Our prototype code in Chap. 3 however does not solve this second-order system. It augments the f and keeps track of the velocities explicitly. This way, it can stick to a first-order formulation of the problem, where the velocities follow the acceleration and the positions follow the velocities. If you have N objects, we obtain $2 \cdot 3N = 6N$ first-order ODEs.

Definition 12.4 (*Linear, homogeneous ODEs*) An ODE is *linear* if we can write it down as

$$\partial_t^{(n)} f(t) = \sum_{i=0}^{n-1} a_i \partial_t^{(i)} f(t) + \hat{F}(t).$$

It is furthermore *homogeneous* if $\hat{F}(t) = 0$.

Linearity for ODEs means linear in the derivatives. You can still have a t^2 term somewhere in there, but may not encounter something like $f(t) \cdot \partial_t f(t)$ or $\left(\partial_t^{(3)} f(t)\right)^2$. If we know that we have a linear, homogeneous equation, then we immediately see that any linear combination of two solutions to the ODE yield a new solution. We call this *superposition principle*.

Definition 12.5 (*Autonomous ODEs*) An ODE is *autonomous* if it has the structure

$$F(f(t), \partial_t f(t), \partial_t \partial_t f(t), \ldots, \partial_t^{(n-1)} f(t)) = \partial_t^{(n)} f(t),$$

i.e. if F does not explicitly depend on t.

Autonomous ODEs have two interesting properties that are useful when we debug code. First, they are translational invariant in time. Let's assume we know a solution for a given initial value. If we have another solution (simulation) that runs into this initial value (and derivative for higher-order ODEs) later on, then it continues along the known trajectory: After all, the t does not enter F, so it does not really matter at which point in time we hit a state. The other solution yields the same trajectory though shifted in time.

The second interesting property for us results from the translational invariance: Let's assume you have found a configuration f such that $F(\ldots) = 0$. This means, you have identified a *stationary* f, as the f does not change anymore from hereon. You therefore can argue about particular solutions to ODEs:

12.2 Attractive and Stable Solutions to ODEs

Many ODEs have—for some particular experimental configurations—solutions that we know or that are simple to construct analytically. The radioactive decay for example has a trivial solution $f(t) = 0$. If there is no substance left, nothing can decompose, i.e. $f(t) = 0$ should not change. Another interesting solution is the orbit

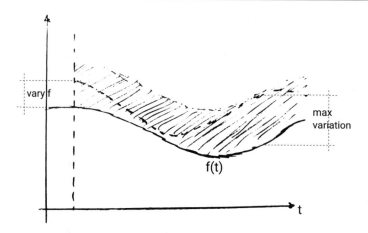

Fig. 12.2 The solution $f(t)$ here is stable: The maximum difference between $f(t)$ and a dotted $\hat{f}(t)$ above is the initial difference times a scaling. So the trajectory of $\hat{f}(t)$ starts slightly off $f(t)$ but then remains within a certain corridor

of a satellite around a planet—you can look up the analytical formula for it. If we start on this orbit with exactly the right velocity, then we should stay on the orbit. By means of its radius, the orbit yields a stationary solution. Obviously, the velocity is not constant as we rotate around the planet. The ODE

$$\partial_t f(x) = \alpha \left(f(x)(\beta - f(x)) \right) \tag{12.2}$$

from Sect. 8.2.2, the *logistic growth*, serves as my last example. Here, $f(t) = 0$ and $f(t) = \beta$ yield stationary solutions. Once we have found trajectories (solutions) of interest for particular setups, we can characterise them:

Definition 12.6 (*Stable solutions of ODEs*) A solution $f(x)$ to an ODE $\partial_t f(x) = F(t, f(x))$ is *stable*, if there is a C such that

$$\forall t, \hat{f}(0) : \qquad |f(t) - \hat{f}(t)| \leq C|f(0) - \hat{f}(0)|.$$

This definition describes the following: If we have a stable solution and wobble around with the initial value, then the resulting new solution doesn't move away from the stable solution arbitrarily. It always remains in a corridor, parameterised by C, around the stable solution (Fig. 12.2).

Stable solution of an ODE

The definition of a stable ODE solution makes a statement about the ODE. It has nothing to do (directly) with the stability statements we make for round-off errors and or time discretisations. These are all different things.

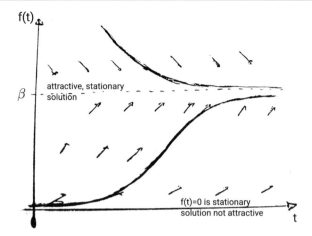

Fig. 12.3 Example visualisation of two solutions for two different initial values of the logistic growth equation (12.2). There are two stationary solution, but only one is attractive/stable. The arrows illustrate the gradients. They make up a *direction field*. Though $f(t) = \beta$ is stable, not the whole environment around it is attractive/stable. Once we go below $f(t) \geq 0$, we have no attractive behaviour anymore. Notably, $f(t) = 0$ is a repulsive, stationary solution

However, if we study the definition closely, we see that the definition has exactly the same structure and format in both cases. This makes sense; otherwise we should have used a different term and not *stability*. And they are linked: If a solution to an ODE is stable, then truncation errors, e.g., put the solution off. But the trajectory of the computed solution does not yield something completely different, unless all the truncation errors all point in the same direction and accumulate. Notably, the underlying problem is well-posed.

Before we discuss this, lets introduce another characterisation:

Definition 12.7 (*Attractive solutions*) A solution $f(t)$ to an ODE is *attractive* if

$$\exists C > 0 : \quad |\hat{f}(0) - f(0)| < C \quad \Rightarrow \quad \lim_{t \mapsto \infty} |\hat{f}(t) - f(t)| = 0.$$

Attractive means that the solution $\hat{f}(t)$ approaches $f(t)$ over time. This can be a very slow process, but they get closer and closer and closer. We immediately see that this tells use something about the condition number: Small alterations of the input data yield no significant long-term difference in the solution. The setup is well-conditioned around $f(t)$. Even better: Any small error we make will be damped out completely on the long run.

If you have a solution, you can formally argue about its attractiveness via some operator eigenvalues. We use a more hands-on approach which usually is sufficient for the kind of problems we study with our codes. It focuses on stationary solutions (Fig. 12.3): Assume you have found one. Move a little bit up and compute the derivative $\partial_t f(t)$—which is easy as it is explicitly given by the ODE. Look only at the derivative's sign. If it is negative, the derivative pushes any solution that is slightly too big down towards the stable solution below. Next, we study trajectories starting slightly below the stable solution and analyse whether $F(t, f(t)) > 0$. If this is the case, starting slightly below the stable solution yields a trajectory which approaches the stable one.

If both tests yield a "yes, it does approach", we have an attractive stable solution. If one of them pushes the off-solution towards the stable trajectory but the other one does not, we have identified a saddle point and the overall thing is not stable: Errors towards the "wrong" side make the trajectory shoot off; errors towards the other side will bounce back, but then we might next shoot in the wrong direction and we are lost again. If neither side approaches, it speak of a repulsive stable solution. Numerical codes basically never hit them.

Once we do our hands-on analysis for our logistic growth equation, we observe a second interesting property: Sometimes, there is a region around a stable solution where other solutions approach it. Here, the solution is stable plus attractive. However, if we start off too far away, then the resulting trajectory is definitely not attractive anymore (Fig. 12.3). Stability and attractiveness can be localised properties: There might be *stability regions* for which everything is fine, but once we leave these regions, anything can happen. Attractive means a special case of stable, but this "attractive implies stable" statement is a property which holds only for scalar ODEs.

Exploit what you know about an ODE

Often, we intuitively know something about our problem when we write code. In an N-body simulation with large N, we can for example not be sure what to expect. However, we can simulate a few simplistic setups first, so we can at least trust our code: Attractive, stable solutions (collisions) should be preserved by all numerical schemes: Even if we suffer from round-off errors and input data errors, they should not blow up.

12.3   Approximation of ODEs Through Taylor Expansion

With our knowledge of ODEs and Taylor refreshed, it is time to revise how we construct the algorithm in Chap. 3. Instead of the complex N-body setup, let the simple ODE

$$\partial_t f(t) = \alpha f(t) \tag{12.3}$$

be our equation of interest. It is supplemented by suitable initial value conditions $f(0) > 0$ and some $\alpha < 0$, so the result will be the decay equation (with the same analytic solution). It yields a curve that approaches $f(t) = 0$ from above.

Definition 12.8 (*Dahlquist test equation*)

$$\partial_t f(t) = \alpha f(t) \qquad \text{with } f(0) > 0 \text{ and } \alpha < 0$$

is called *Dahlquist test equation* or only test equation. It serves us for many purposes, as it is simple yet not trivial and we know its analytical solution.

The test equation's analytical solution is infinitely smooth, i.e. you can throw ∂_t on it infinitely often. None of its derivatives does disappear. As the function is that smooth, we know that the Taylor expansion works well. If we have infinitely many terms in it, then it will *exactly* yield the solution.

As we can not handle sums of infinite length on a computer, we truncate it. This is one fundamental idea behind many approaches in the simulation business. We take the Taylor expansion of the solution

$$f(t + dt) = f(t) + dt \cdot \partial_t f(t) + \underbrace{\sum_{i \geq 2} \frac{dt^i}{i!} \partial_t^{(i)} f(t)}_{\text{higher order terms}}$$

and throw away the higher-order terms. Next, we take the $\partial_t f(t)$ term and plug in the ODE as written down in first-order formulation.

The result is a time-stepping scheme, i.e. it sketches how we can write down a for-loop running through time with a step length of dt. Since we know how to rewrite any system of ODEs of any order into a first-order formulation, and as the truncation idea works for both scalar- and vector-valued $f(t)$, we can solve any ODE via this scheme.

Definition 12.9 (*Explicit Euler*) Take the solution and develop its Taylor expansion. Truncate this expansion after the second term and plug the ODE definition into the derivative. This yields the time stepping loop from the model problem. It is called *explicit Euler*.

Some Important Material

- B. Aulbach: *Gewöhnliche Differentialgleichungen*. Spektrum Akademischer Verlag) (1997) (in German, but maybe the best introductory book for ODEs)

> **Key points and lessons learned**

- The term ODE is well-defined as well as a lot of terminology around ODEs.
- It is clear how to rewrite a higher-order ODE into a first-order system of ODEs.

- There are three properties for ODE solutions which are important for us: stationary, stable, attractive. If we solve a problem with our code, it is reasonable first to validate our code against (artificial) scenarios for which we know the solution or the solution properties. The tests help us to find bugs, and they allow us to quantify (if we know an analytical solution) how bad our computer's approximation is.
- The explicit Euler is a straightforward outcome of a truncated Taylor expansion with only two terms into which we plug the ODE.

Accuracy and Appropriateness of Numerical Schemes

13

Abstract

For the accuracy and stability of numerical codes, we have to ensure that a discretisation is consistent and is stable. Both depend on the truncation error, while we distinguish zero- and A-stability. We end up with the notion of convergence according to the Lax Equivalence Theorem, and finally discuss how we can compute the convergence order experimentally.

We study two particles as they orbit around each other. They dance through space. Intuitively, we understand that this dancing behaviour should become the more accurate the smaller the time steps. In the sketch below, the top "animation" runs with half the time step size compared to the bottom simulation.

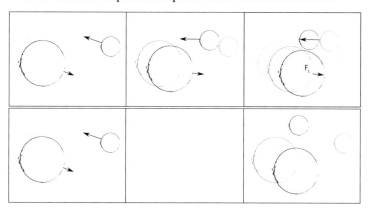

The left (starting) frame is exactly the same. It hosts two particles with two velocities. The bottom row performs one time step, whenever the top row performs two time steps with half the time step size. Due to the more of time steps, the top row yields a different, more accurate particle configuration in the right frame.

Accuracy is what we want to quantify now. Unfortunately, we lack a clear reference point: We don't know where the particles should (analytically) be after a certain time T. Therefore, we track the positions of the particles $p_1(T)$ and $p_2(T)$ after a certain fixed time T and compare them with each other. What do you expect if you simulate the time span $(0, T)$ with a time step size dt, then with $dt/2$, then with $dt/4$, …?

We derive our explicit Euler from the Taylor expansion by truncating terms:

$$f(t + dt) = f(t) + \underbrace{dt \cdot \partial_t f(t)}_{\text{plug in F}} + \underbrace{\sum_{i=2}^{\infty} \frac{dt^i}{i!} \partial_t^{(i)} f(t)}_{\text{cut off}}.$$

In this formula, we assume that the Taylor series for $f(t)$ exists within a dt-area around t (cmp. criteria on page 134). Therefore, the sum over i quantifies the error that we introduce by solving the problem numerically instead of analytically:

Definition 13.1 (*Truncation error*) The *truncation error* is the error that results from the time discretisation, i.e. from cutting off the Taylor series.

Sampling a phenomenon in intervals dt is something we do all the time in scientific computing, as we have to break down the problem over t into a finite number of calculations with finite cost for our machines. Scientists often use the letter h for the sampling size or step size. While we have used dt for time step sizes so far, we will often use h for this discretisation width from hereon. h describes the time in-between two time samples. When I write $f_h(t)$, the subscript h highlights that f is defined over a continuous t which is sampled with finite step size.

We make two types of errors with our approach: we face round-off errors introduced by floating point precision, and we truncate the Taylor series which allows us to discretise the time. The round-off errors, i.e. arithmetic stability, are something we have discussed in Chap. 8. Let the subscript M highlight that there is some round-off error "hidden" within a value. The subscript h emphasises that we make a discretisation error. In practice, the two errors team up. Often, books make h include the round-off error M, too, as the truncation error typically is the dominant one.

Definition 13.2 (*Consistency*) A *consistent* numerical scheme is a numerical scheme that solves the *correct* problem as its discretisation step size goes to zero.

A more formal definition goes as follows: Let τ be the truncation error, i.e. the stuff we cut off. Consistency means

$$\lim_{h \to 0} \frac{\tau}{h} = 0,$$

i.e. the truncation error goes down faster than h.

Consistency in linear algebra

People use a slightly alternative formalism if they have been brought up with linear algebra: If you solve $Ax = b$, then an algorithm is called *consistent* if it yields— in exact arithmetics—the correct solution for any b. This seems to be a different notion of consistency, but it is actually not: If we write down the explicit Euler as a matrix-vector product, we obtain

$$id \cdot f(t + h) = f(t) + B(h)f(t) = (id + B(h))f(t). \qquad (13.1)$$

The $f(t)$ is the vector of particle velocities, and the $B(h)$ is a matrix encoding the force impact times dt, while id is the identity matrix. In other words, we solve an equation system $Ax = b$ per time step with a trivial matrix $A = id$ serving as left-hand side. The right-hand side equals $(id + B(h))f(t)$. $f(t+h)$ consequently plays x's role. According to our ODE terminology, this scheme is consistent if it yields the continuous ODE's solution for $h \mapsto 0$; which is equivalent to getting the correct solution of (13.1) even for the limit $b = \lim_{h \to 0}(1 + B(h))f(t)$ and for any $f(t)$.

For the explicit Euler, we directly see that our approach is consistent. Once we have written down the scheme, we replace the $f'_h(t)$ in the Taylor series with the given ODE definition

$$f'(t) = F(t, f(t)),$$

and argue that the remaining terms fade away with h^2 once h becomes small. That is faster than h. We solve *the right problem* when h goes to zero. For the explicit Euler, the claim that we are consistent is trivial.

Non-consistent schemes

More sophisticated numerical schemes that approximate higher-order terms in the Taylor series to yield better estimates are *not* automatically consistent. In this case, we have to invest work to show that we solve the correct problem as $h \mapsto 0$.

We assume that a problem is well-conditioned and that our numerical scheme is consistent. As the truncation error teams up with the round-off error, we ask ourselves two questions:

- Is the solution also stable for any dt (or h; whichever notation we do prefer), i.e. do we always get something meaningful?

- How do the truncation and the round-off errors sum up, and what is the accuracy we obtain in the end?

When we focus on the first question, we basically ask our consistency questions the other way round: We know that we get the right answer for $dt \mapsto 0$. But what happens for bigger dt?

13.1 Stability

Before we study complex challenges such as our multibody simulation and before we formalise yet another notion of stability, we study what could possibly go wrong with an explicit Euler. For this, it makes sense to focus on the Dahlquist test equation. Writing down the explicit Euler for this setup is reasonably simple, as we can plug in the right-hand side F and isolate the $f_h(t + h)$:

$$f'(t) = \lambda f(t) \quad \text{with } f(t = 0) = f_0 \text{ is the continuous model.}$$
$$f_h(t + h) = f_h(t) + h \cdot f_h'(t) = f_h(t) + h\lambda f_h(t) = (1 + h\lambda) f_h(t), \text{ i.e.}$$
$$f_h(t + Nh) = (1 + h\lambda)^N f_h(t). \tag{13.2}$$

We take our explicit Euler formula and plug the $F(t, f)$ in. It is, in this case, only an $F(f)$ as the ODE is autonomous. We then bring the $f_h(t + h)$ to the left side and all the other terms to the right. For the test equation, this works. For "real" problems, it is often not possible (straightforwardly). As the explicit formula for a time step is a multiplication of the previous solution with a weight, we can now iteratively apply this scheme. Eventually, we directly compute the Nth time step from $f(0)$ without constructing the intermediate steps.[1] Again, this is usually not that simple for more complex F.

The following code snippet computes the explicit Euler including the intermediate values, i.e. it constructs the solution step by step. It exploits the fact that we know an analytical solution for this particular problem and determines not only the solution but also the error per step:

```
double f0     = 4;       // initial value
double t      = 0.0;     // time
double lambda = -1.0;    // ODE parameter (fixed)
double h      = 0.001;   // time step size
double T      = 2.0;     // terminal time
double f      = f0;
```

---

[1] N here is a counter for the time steps. It has nothing to do with the N from "N-body simulations".

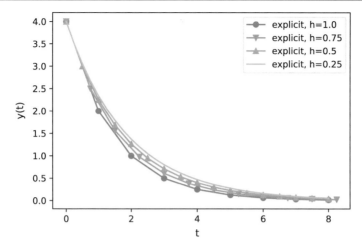

Fig. 13.1 Explicit Euler trajectories for the Dahlquist test equation with $\lambda = -1$ and various time step step sizes

```
double e      = 0.0;    // error
while (t<T) {
    t += h;
    f = (1+lambda * h) * f;
    e = f - f0 * exp(lambda * t);
}
```

When we dump the outcome for smaller and smaller h, we observe interesting things:

- For $h = 0.001$ there is a significant round-off error in t. As we can not represent 0.001 exactly in floating point precision, the h will be slightly bigger or smaller than our value of choice. The resulting error accumulates and we thus overshoot T: In one step, t is just slightly smaller than T. In the next step, it is bigger.
- The smaller we make h, the smaller the final error at $t \approx T$. This is not a surprise. Our scheme is consistent and we expect the round-off errors to disappear with smaller hs, too.

A plot of the behaviour for various time step sizes confirms this conclusion (Fig. 13.1): With $h = 1.0$, we are far away from the analytical solution of the test equation. Once we make the h smaller, we get closer and closer to the real behaviour. We also observe that we undershoot the real solution (cmp. to the sign in the above error computation) all the time: The explicit Euler starts from a point and then walks along the derivative for a time interval of h. Since the function smooths out, i.e. the derivative approaches 0, we always go down too aggressively.

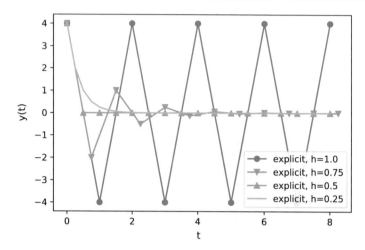

Fig. 13.2 Explicit Euler trajectories for Dahlquist test equation with $\lambda = -2$ and the same time step step sizes as in Fig. 13.1

We next decrease the material parameter λ.[2] We check $\lambda = -2$ (Fig. 13.2). While we see a similar behaviour as before for very small time step sizes, we get oscillations for bigger time step sizes. These oscillations damp out as long as $h < 1$. For $h = 1$, there is no damping at all and the solution does not approach the real solution of $f(t) = 0$ for large t. If $h > 1$, the amplitudes grow.

Everything seems to be right if we have a smallish time step size. For $h = 0.5$, we run along the gradient "too long", end up exactly on $f_h = 0$ which is an attractive stable solution, and thus remain on the axis. Trivially, our numerical solution on the long run is consistent, but we struggle to call the observed curve characteristic for this problem. For h slightly bigger than 0.5, we run for too long along the derivative in the first step and obtain an approximate value $f_h < 0$. As a consequence, the derivative becomes positive, i.e. switches the sign, and we go up again. We produce an oscillation. The code still yields the right solution on the long term (the solution goes to zero), but this is qualitatively yet another solution than the ones we have had before. Even worse is the situation for $h > 1$. We create oscillations where the amplitude becomes bigger and bigger.

Definition 13.3 (*Zero-stable*) A numerical scheme is *zero-stable* if the solution goes to zero for the Dahlquist test equation with $\lambda < 0$ for $h \mapsto 0$.

The explicit Euler is zero-stable. Zero-stability is essential to have a consistent scheme. As we also know that the round-off errors of the explicit Euler are kind

---

[2] I call λ the material parameter. Different to flow through a subsurface medium, e.g., our λ is not really a material. But the name material here highlights that it is not a parameter determined by yet another equation but something fixed.

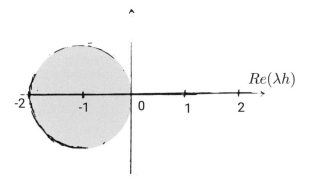

Fig. 13.3 We assume that $h\lambda$ of the Dahlquist test equation spans the complex plane. So the x-axis is the real part (which we usually use, but this plot makes no constraint on λ, i.e. allows it to be complex-valued). Only if the product $h \cdot \lambda$ ends up in the shaded circle, the numerical scheme is stable. The calibration of $h\lambda$ seems to be weird: We would maybe expect only h on the x-axis. However, with h on the x-axis, we would need a different plot per λ. The present plot holds for all λ. It clarifies which h yield a stable solver for all λ

of under control, i.e. as we know that we have the implementation error under control, we do not really have to worry for $h \mapsto 0$. However …

Definition 13.4 (*A-stable*) The *A-stable* region of a numerical scheme for the test equation is the set of (h, λ) parameters for which the scheme converges towards zero. A numerical scheme is A-stable if it is A-stable for all (h, λ) inputs with $\lambda < 0$.

As we see above, the explicit Euler is not A-stable. To formalise this behaviour, we go back to (13.2) where we write down the explicit formula computing the outcome for the test equation after N time steps in one rush. To ensure

$$f_h(Nh) = (1 + h\lambda)^N f_h(0) \mapsto 0,$$

we need $-1 < 1 + h\lambda < 1$. The whole study has to be done for $\lambda < 0$ (we want the analytical solution to decrease towards zero) and we know that h is positive (we can not go back in time). As a consequence, $1 + h\lambda < 1$ is trivial. We however have to be concerned about $-1 < 1 + h\lambda$ which implies

$$0 < h < -2/\lambda. \tag{13.3}$$

This inequality determines our A-stable region.

Scientists like schematic illustrations. Figure 13.3 introduces the standard way to depict the insight from (13.3): If $\lambda > 0$, we are lost right from the start. If $\lambda < 0$, we have to meet (13.3). In theory, $\lambda \in \mathbb{C}$. We thus care about the magnitude of λh. If $\lambda \in \mathbb{R}$ (which is what we have in this manuscript), we are only interested in the x-axis.

> **Relation to N-body code**

If two objects are reasonable far away from each other, the above observations tell us that we are basically fine. Once two objects however get close, we know that we have to be extremely careful with the time step size, as the forces "explode" and thus yield a massive position and velocity gradient. This is the N-body counterpart of very small $\lambda \ll 0$ in the Dahlquist test equation. Indeed, we can bring two objects "arbitrarily" close and thus force ourselves to use smaller and smaller time step sizes. This behaviour is later formalised under the label *stiff equation*.

13.2 Convergence

With our notion of the stability and consistency, we can finally bring both concepts together:

Definition 13.5 (*Convergence*) An numerical scheme for an ODE *converges* if

$$\lim_{h \to 0} \max_N \| f_h(Nh) - f(Nh) \| = 0.$$

Convergence means that no matter at which time step N we study a solution, we close in on the right answer once h goes to zero. We can phrase the definition differently:

> **Lax Equivalence Theorem**

A numerical scheme converges if and only if (iff) it is consistent and zero-stable.

I prefer the Lax Equivalence Theorem as it highlights explicitly that algorithm choice and implementation go hand in hand: When h approaches zero, we want to be sure that we solve the right problem plus that no round-off error pollutes the outcome. In accordance with the definition, we can now also speak of convergent codes. They are stable realisations of a consistent numerical scheme. Notably, consistency and stability are both necessary for convergence, but each property alone is not sufficient.

Terminology

Convergence is often used in a sloppy way and some people use it as synonym for "nothing changes in my iterative code anymore". Always check whether it includes both stability and consistency. You might for example have run into the wrong solution if a while loop terminates as it does not change the solution anymore, or you might miss the right solution as you decrease h since your round-off errors mess everything up.

In practice, we can not show real convergence (or consistency) in line with the mathematical definition. Once h ends up in the order of machine precision, the error will likely plateau as well. Yet, a decent selection of small h values that exhibit a clear trend is usually seen as sufficient evidence—despite the fact that even smaller h might introduce compute instabilities.

So far, our convergence statement is a qualitative one. We next derive a quantitative notion of convergence. For this, we study the truncation error we make: We compare the numerical approach $f_h(h)$ with the not-truncated Taylor series of $f(h)$, i.e. the real solution, for one time step of h. This gives us

$$
\begin{aligned}
e_h(h) &= f_h(h) - f(h) \\
&= f_h(0) + hf'(t, f_h(0)) \\
&\quad - \left(f(0) + hf'(t, f(0)) + h^2/2 f''(t, f(0)) + h^3/3! f'''(t, f(0)) + \ldots \right) \\
&= f_h(0) - f(0) + h\left(f_h'(t, f_h(0)) - f'(t, f(0)) \right) + h^2/2 \left(-f''(t, f(0)) \right) + \ldots \\
&= -h^2/2 f''(t, f(0)) + \ldots
\end{aligned}
$$

In this ansatz, we develop the solution at $t = h$ from the initial condition $t = 0$. We plug in our assumption/knowledge that the initial condition is the same for both the numerical scheme and the real solution. $f_h(0) = f(0)$. At this point, we ignore that we know that $f(0)$ is slightly wrong. We assume that our setup is well-conditioned—otherwise we would not have started to think about convergence. If $f_h(0) = f(0)$, then $f_h'(0, f_h(0)) = f'(0, f(0))$, if we study consistency, or $|f_h'(0, f_h(0)) - f'(0, f(0))| \leq C\epsilon$ subject to an additional h-scaling otherwise. With $\epsilon \cdot h \approx 0$, the second term drops out, too. We end up with a remainder that is scaled with h^2.

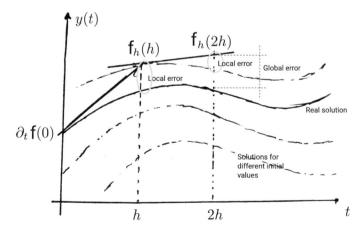

Fig. 13.4 Illustration of the global and local error. As the real solution meanders up and down, we observe that our global error statements are a worst-case estimate. Sometimes we overshoot, sometimes we undershoot. The local error is the sum of local errors (though each local error corresponds to a different ODE solution) which might cancel out

Since $f''(t, f(0))$ is given by the problem, i.e. it is a real "constant" in this sense, we are zero-stable. The error disappears for $h \mapsto 0$. Zero-stability is no surprise, but we now can quantify the error:

Definition 13.6 (*Convergence order*) The *convergence order* p of a numerical scheme is given by

$$|e(t)| \leq Ch^p.$$

The local truncation error of the explicit Euler is in $\mathcal{O}(h^2)$. The explicit Euler's single step is thus second order.

However, this is only the analysis of the first step. Once we have completed the first step, i.e. moved from the initial condition to $f_h(h)$, the second term in the error computation $f_h'(t, f_h) - f'(t, f)$ does not vanish anymore. Therefore, we carefully have to distinguish a local error (made by one time step) vs. the global error that we actually observe. They are two different things (Fig. 13.4), as the reference point for both quantities differs: it is either the starting point of a time step (local error) or the real solution (global). The global error is an accumulation of local errors; although, in practice, local errors often cancel out as we have seen it before for the floating point numbers. This is the same story as for the truncation error, i.e. all just a little bit of history repeating!

Convergence vs. consistency order

A more formal script would distinguish consistency order and convergence order. Convergence order takes the implementation into account, too, whereas consistency assumes exact arithmetics. In practice, we always measure the convergence order which is dominated either by the consistency order or arithmetic errors—whichever is bigger.

Deriving the global error requires a little bit more work than the local error. Let's assume that we run our algorithm over a time span T, i.e. that we do $\frac{T}{h}$ time steps. We start with an analysis of the last step of our scheme which suffers from the error

$$|e_h(T)| = |f_h(T) - f(T)|$$

$$= |\underbrace{f_h(T - h) + hF(T - h, f_h(T - h))}_{\text{one Euler step}} - f(T)|$$

$$= |f_h(T - h) + hF(T - h, f_h(T - h)) - \left(f(T - h) + hF(T - h, f(T - h)) + \frac{h^2}{2} \ldots \right)|$$

$$\leq |f_h(T - h) - f(T - h)| + h \cdot |F(T - h, f_h(T - h)) - F(T - h, f(T - h))| + h^2|\frac{1}{2!} \ldots|$$

$$= |e(T - h)| + h \cdot |F(T - h, f_h(T - h)) - F(T - h, f(T - h))| + h^2|\frac{1}{2!} \ldots|$$

While we have used the same ingredients as before—our definition of an explicit Euler plus the Taylor expansion—the important difference is that the $F(T - h, f_h(T - h))$ in the Euler part is fed the numerical solution $f_h(T - h)$ from the previous time step. It does not disappear. Instead, we keep it and apply the triangle inequality to the whole expression. I omitted the absolute values for the error before to highlight that the error can be a "too much" or "not enough", but now we need the norm. We obtain a recursive expression for the error, where the error (bound) increases per time step by the inaccuracies in the first derivative plus an additional $\mathcal{O}(h^2)$ term that is also often called *global error increment*.

To get the second term under control, we exploit the fact that our ODEs have to be *Lipschitz continuous*.

Definition 13.7 (*Lipschitz continuous*) Let f_1 and f_2 be two solutions to the ODE
$$\partial_t f(t) = F(t, f(t)),$$
i.e. f_1 and f_2 belong to two different initial values. The F is *globally Lipschitz-continuous* iff
$$|F(t, f_1(t)) - F(t, f_2(t))| \le C|f_1(t) - f_2(t)| \quad \forall t \ge 0.$$

Indeed, one can show that this is the only condition we need to ensure that the ODE has a solution. We therefore can safely rely on the fact that this holds for our problems of interest. While the proof is out-of-scope, we do emphasise what it means: If we slightly modify the solution f, then the F-value does not change dramatically. It means that we are kind of well-conditioned along the $f(t)$-axis. Plugging in a "wrong" f into our solver will yield an error, but this error will not be amplified arbitrarily.[3]

Lets use the continuity property in our global error estimate. We obtain

$$|e_h(T)| \le |e(T-h)| + h \cdot |F(T-h, f_h(T-h)) - F(T-h, f(T-h))| + h^2|\frac{1}{2!}\ldots|$$

$$\le |e(T-h)| + h \cdot C|f_h(T-h) - f(T-h)| + h^2|\frac{1}{2!}\ldots|$$

$$= (1 + hC)|e(T-h)| + h^2|\frac{1}{2!}\ldots|$$

where the C is the so-called *Lipschitz*-constant from the definition. Some remarks on this formula:

- One might assume that the explicit Euler does not converge, as the recursive equation above makes us use $\frac{T}{h}$ steps. In each step, we get an additional error amplification from the left part of the right-hand side in the order of h. So it seems that we obtain $\frac{T}{h} \cdot h$. However, the initial error is zero! Therefore, if we unfold the $(1 + hC)|e(T-h)|$ term, we eventually hit $e(0) = 0$ and the term drops out.

---

[3] Some maths books define Lipschitz-continuity "simply" as $|F(s_1) - F(s_2)| \le C|s_1 - s_2|$. In our discussion, we split up this s into $s = (t, f(t))$ as we are interested in ODEs, and we wobble around with the $f(t)$ part only. The more general definition from math books shows that we also can slightly alter the t argument. The solution will not change too much either. Both definitions focus on the right-hand side of the ODE. As the right side determines the solution, its (continuity) properties carry over to the solution.

- Each recursive unfolding step (induction or time step) produces one $h^2 |\frac{1}{2!} \ldots|$ term. In the end, we will have N of these guys. So their number scales with T/h or $\mathcal{O}(h^{-1})$.
- These remaining terms scale with the second derivative $\partial_t^{(2)} f(t)$. It is the inaccurate capture of the function's second derivative that injects more and more error into the approximation.

The interested student can now study the discrete Gronwall lemma. For us, we set $N = T/h$ and write

$$|e_h(T)| \leq \underbrace{(1 + hC)|e(T - h)|}_{unfold} + h^2 |\frac{1}{2!} \ldots \partial_t^{(2)} f(t)| + \ldots$$

$$\leq \underbrace{(1 + hC)^N |e(0)|}_{=0} + \underbrace{N \cdot h^2 |\frac{1}{2!} \ldots \partial_t^{(2)} f(t)|}_{\text{recursion spills out terms}} + \underbrace{\ldots}_{\text{ignore higher order terms}}$$

$$\leq hC \cdot h^2 |\frac{1}{2!} \ldots \partial_t^{(2)} f(t)|$$

$$\leq Ch |\max \partial_t^{(2)} f(t)|.$$

The "ignore" remark means that the magic constant before the preceding term has to be increased accordingly.

> **Convergence order explicit Euler**

The explicit Euler converges linearly.

This is the *global error statement*: If we half the time step size h, we can also expect the error to go down by a factor of two.

13.3 Convergence Plots

One challenge we face for real problems is that we typically can not compute the error, as we do not know the exact solution. In this case, it helps to compare two errors for different mesh sizes:

$$e_h(T) - e_{h/2}(T) = (f_h(T) - f(T)) - (f_{h/2}(T) - f(T))$$

$$= \underbrace{f_h(T) - f_{h/2}(T)}_{\text{measure}} = Ch^p - C(h/2)^p = C(1 - \frac{1}{2^p})h^p.$$

We measure the outcome for two different time step sizes and plug in our model of a convergence order (Fig. 13.5). This yields one equation for two quantities C and

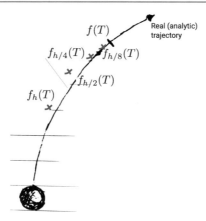

Fig. 13.5 A planet runs through space on a trajectory which is unknown to us. To study the convergence behaviour of our code, we can however fix a certain observation time T. We still do not know the solution $f(T)$, but we know that the points $f_h(T)$, $f_{h/2}(T)$, $f_{h/4}(T)$, ... close in on the solution for a consistent algorithm

p. If we repeat this for the next pair of measurements

$$e_{h/2}(T) - e_{h/4}(T) = f_{h/2}(T) - f_{h/4}(T) = C(\frac{1}{2^p} - \frac{1}{4^p})h^p = C(1 - \frac{1}{2^p})\left(\frac{h}{2}\right)^p,$$

we see that the difference between two subsequent measurements goes down in line with the order.

To determine the convergence order of a scheme for a given setup experimentally, we thus fix a quantity of interest f at a certain time T, and then we make a series of measurements. The result data itself is not of primary interest this time. We are interested in the differences between two subsequent measurements for different h, as we fit a curve through the differences to validate our claim about a certain convergence order:

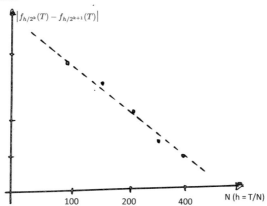

Note that this sketch of a plot has flaws (cmp. remarks in Appendix F). Yet, it illustrates the principle behind our experimental strategy plus the fact that it is very convenient for such plots to use log scales. An interesting decision is that I did not use h on the x-axis. With h on the x-axis, left would would mean accurate ("smaller h is better"). Many readers intuitively associate right with better, so rather than depicting h, I used the time step count as unit. Further right thus means more accurate.

It is never a good idea to study only one setup or only three or four step sizes to claim a certain convergence order. Run more experiments. If all of them yield parallel lines, then you can claim that you observe a certain order. This is in line with our previous observation that stability and consistency can depend on the experimental setup.

Some important material

- Wolfgang Hackbusch: *Iterative Solution of Large Sparse Systems of Equations.* Applied Mathematical Sciences (vol. 95). Springer Verlag (2010)

> **Key points and lessons learned**

- Truncation errors determine how accurate we solve our ODE.
- Convergence means consistency and stability.
- We can quantify the convergence order analytically,
- but we can also determine it experimentally by comparing outcomes from different simulations.
- Researchers distinguish zero-stable from A-stable. The former is important for consistency discussions, while the A-stability is more important for day-to-day simulation runs.

Wrap-up

In this part of the manuscript, we revise some of the most fundamental concepts behind numerical algorithms. Our discussion orbits around our model problem—a few particles "dancing" through space—or around very basic ordinary differential equations. The simplest case is the Dahlquist test equation. Though basic, these examples help us to establish the language we need in almost any area of scientific computing. Terms like consistency, stability or truncation error are found in the field of numerical linear algebra or for the solvers of partial differential equations as well. They are omnipresent.

With our toolset, i.e. an appropriate language and fundamental ideas, we are in a position to write N-body simulations and analyse their outcome, plus we can discuss, assess and understand their characteristics. Our work is not mere code-monkeying anymore, since we know why we are writing certain code, how accurate it is, and when and why it breaks down. This toolset also maneuvers us into a position where we can introduce more sophisticated numerical algorithms.

While the latter is among our long-term goals, this manuscript departs from the route of numerical algorithms for the time being, and instead equips the reader with the skill set and knowledge to write simulation code that makes use of today's omnipresent multicore computers. We also introduce tools to employ GPGPUs. Overall, our intention is to become able to write simulations with really large particle counts that run over a longer time span. In my opinion, it is reasonable to discuss better numerics with a large-scale mindset. Designing mathematically advanced algorithms is a challenge. Designing mathematically advanced algorithms that scale is a harder challenge. As the basics of numerics are now known to us, we next acquire a sound understanding how parallel codes can be written.

Part V

Using a Multicore Computer

Writing Parallel Code

14

Abstract

We distinguish three different strategies (competition, data decomposition, functional decomposition) when we design parallelisation schemes. SIMD's vectorisation is a special case of data parallelism which we now generalise to MIMD machines where we distinguish (logically) shared and distributed memory. Our text discusses some vendor jargon regarding hardware, before we introduce BSP and SPMD as programming techniques.

Assume there are p boxes filled with Lego© bricks of p different colours. We label them as input boxes. Every box holds a random number of bricks and colours. Let's further assume that you have a team of p colleagues, and each one "owns" one of the filled (input) boxes. Finally, each colleague has one box of its own which is initially empty—we call it output box. Colleagues are allowed to chat with each other. They also can exchange input boxes. However, your colleagues can not handle more than one input box at a time, and they can only access their own output box.

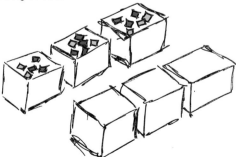

How would you solve the following challenges?

> 1. How many red bricks are in all the input boxes together?
> 2. Is there a red brick in one of the boxes?
> 3. Sort all red bricks into the first output box, all green bricks into the second, and so forth.

In my opinion, there are three fundamentally different ways to come up with a parallel solution for a problem. Obviously, not all problems are a fit to all three approaches, but often they complement each other. Let us assume that a problem decomposes into subproblems, and that we have various compute entities, where entities can mean cores or computers or servers.

1. *Functional decomposition.* The problem decomposes into different steps A, B, C,...Some of these steps—often mapped onto different function calls—can run in parallel. We deploy them to different compute units.
2. *Data decomposition.* This is maybe the most popular parallelisation approach. Assume our task is to run over a large dataset. We can split the dataset into smaller chunks which means we split our task into subtasks over subchunks. We distribute the subproblems among the compute units.
3. *Competition.* We make each unit either handle a different task which searches for a solution and/or handle a different subproblem. The first unit which finds a solution to a given problem tells the others to stop.

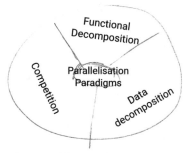

Functional decomposition is something we find among construction teams. The first worker in a car factory builds the chassis, and the second paints it while the third one mounts the wheels. Any gain from parallelism results from different activities that can be completed in parallel.

Data decomposition in this example is if we make four workers mount the four wheels in parallel. The gain stems from the fact that one activity is to be done over multiple items, so we split this up.

SIMD and Data Parallelism

SIMD is an extreme case of data parallelism where the individual steps per work item run in a synchronised way.

Competition is something we might find in search algorithms, e.g. If we search for the best strategy in Chess, we might ask different computers to run through different strategies (search trees). The first one coming up with a winning strategy makes the others stop. If we search for patterns in a signal (from outer space, e.g.), we might split up the data into chunks and ask different computers to process the chunks in parallel. The first one finding signals from an alien is the winner.

Parallelisation by competition is something we will not investigate further in this text. We instead continue with data decomposition—we have already followed this route with SIMD—and then transition into programming techniques for functional decomposition.

Pipelining

Even if steps within a functional decomposition have to run one after another, and even though we might not be able to run steps for different work items in parallel—due to a shortage of well-suited compute components such as vector computing units, e.g.—there is an additional, potential gain from parallelism if we can functionally decompose our work into a series of activities A, B, C,…and have to run over a sequence of problems: While worker number two in the car manufacturing example above starts painting, the first worker already continues with the next car. In the end, we get the impression that cars are built in parallel even though car by car drop out sequentially.

This pattern is called *pipelining*: We give the first subproblem to the first unit running the first task (a particular activity) on this subproblem. Meanwhile, we give the second subproblem to the second unit which performs a different task. Whenever the entities have completed their task, they pass the subproblem on to another unit.

Pipelining (as materialisation of a functional decomposition) and data decomposition can often be applied as alternatives or can be combined. Pipelining is typically more complex and always needs some warm-up and cool-down phases (when not all steps of the car manufacturing pipeline are occupied, e.g.). It is also notoriously vulnerable to a step within the pipeline that lasts longer than all others; the slowest step determines the throughput.

Pipelining is heavily used by hardware. If a particular instruction runs through S phases in a pipeline—phases typically are "understand the type of operation", "bring in the variables required", "write the outcome into a register", …—the pipelining concept can bring down the effective cost per instruction to 1: In the ideal case, one instruction "leaves" the chip per second. This obviously implies that instructions are reasonably independent of each other such that we can progress them independently. In this ideal case, it seems that S steps are done per time unit. From a user's perspective, we may thus argue that we have done multiple instructions for one piece of data in parallel. The acronym in Flynn's taxonomy for this is MISD (*Multiple Instructions Single Data*).

14.1 MIMD in Flynn's Taxonomy

Before we study data and functional decomposition further, we introduce a more flexible machine than our SIMD kit: a multicore computer. They are comprised of multiple instances of the von Neumann architecture that all sit on one node (computer) and share the memory. That means that there are multiple independent minicomputers on our chip. Each one has its own program counter and ALU. Each core can work on something completely different, independently from the others. Each core handles its instructions in its own pace. There is no (built-in) synchronisation how to run through the instruction stream (program).

Definition 14.1 (*Multiple Instructions Multiple Data*) A machine with multiple compute units that run through their own instruction sequence and operate on data of their own is classified as MIMD (*multiple instruction multiple data*) architecture.

The above definition doesn't prohibit that two compute units share data. It just means that everyone works on its piece of data at a time (in the registers). Nevertheless, these "multiple data" are all allowed to reside within the same memory.

Definition 14.2 (*Shared memory versus distributed memory*) A *shared memory* machine is a machine where multiple compute units share a common data region, i.e. can access the same memory. In a *distributed memory* environment, the memory per compute unit is not connected, so if one compute unit wants to access data of the other one, the two units have to exchange the data explicitly via a message.

MIMD is a hardware classification. Both multicore computers and whole clusters are materialisations of this paradigm. ALUs with vector capabilities however are not. They run the same instruction per data tuple even though these instructions might be subject to masking.

Shared Versus Distributed Memory Programming

The distinction of shared versus distributed memory can be a logical or a technical one: You can start two applications on your local laptop (you do so all the time when you open an email client and a PDF viewer, e.g.). These two codes do not share any data even though their data reside in shared memory. Logically, we work with a distributed memory setup. Technically, we operate with shared memory. There are hardware examples and software abstraction layers for the other way round, too.

In practice, hardware terminology around shared multicore architectures quickly becomes confusing; and many vendors add their bits and pieces to make the zoo of terms bigger. Here are some frequently found terms:

1. The von Neumann pair of control unit plus ALU (incl. an FPU if required) is a *core*. So each core plus bus plus memory yields a complete von Neumann architecture, but bus and memory are shared between cores.

2. Some vendors speak of (hardware) thread rather than core. At the same time, we call a thread in a programming language one execution stream running independently of the other threads. So if someone tells you about a thread, this might mean a (software) thread where a piece of the program runs its code in a MIMD sense (though MIMD in turn is a hardware term) or a real thing in hardware.

3. You can run multiple logical (program) threads, i.e. different instruction pathways, on one core. It is the responsibility of your operating system to swap them. Actually, you do this all the time on your laptop when you open Skype, emails, Word, Twitter, …Obviously, if multiple threads are executed on one core, we cannot expect any improvement in runtime. Unless there are enough registers around to store data temporarily—that's what GPUs do—there will be a high price to pay for the swaps. For our purposes, it usually does not make sense to have more threads than CPU cores.

4. On GPGPUs, it makes sense to have way more logical than physical threads (hardware),[1] as GPUs have a massive number of registers. So they do not have to empty and free registers into the memory if another thread continues. They simply make the other thread use another set of registers. The hardware threads on a GPU are also sometimes called CUDA threads.

5. Today, most vendors ship chips with *simultaneous multithreading (SMT)*. If your chip has SMT, it means that a core has N_{SMT} hardware threads. However, these (hardware) threads share some parts of the chip such as the FPU. So if one of the threads runs into some floating point calculations, it occupies the FPU, and the other threads have to wait if they need the FPU, too. Intel brands its SMT with $N_{SMT} = 2$ as *hyperthreading*. The machine or the sales department of the vendor, respectively, pretend to give you N_{SMT} full cores/threads, but they are actually not. Through this trick, most systems today seem to have twice (or even more) as many threads as full-blown cores. For most standard (office) software, the lack of parallel FPU silicon is not a big deal, but for scientific software, we have to carefully evaluate the role of hyperthreading. It might be "fake multithreading".

6. The majority of compute servers today are shipped in a two-socket configuration. A socket is a piece of silicon that hosts several cores and their shared memory. If you have two sockets, the two sockets connect their memory and thus pretend to have one large shared memory serving twice the core count (cmp. tips in Sect. C.2). In reality, these are however different pieces of memory wired to each other. They are not really one homogeneous thing. This means, you might face runtime penalties (the system has to ship data over this wire) when one socket needs memory unfortunately stored on the other guy. In this introductory text, we do not really care about the arising non-uniform memory access cost (scientists

---

[1] GPUs in the classic CUDA tradition use a slightly different terminology as discussed in Chap. 18.

speak of *non-uniform memory architectures* (NUMA)), but we have to be able to understand the following terminology when we read a computer spec:

7. A *node* is what a lay person would call the computer. Nodes are connected via a fast interconnect, but they do not share memory. Each node consists of *sockets*, typically two of them. Each *socket* hosts a number of physical *cores*. Each core hosts *hyper-/SMT threads*. Most supercomputers hence offer

$$N_{nodes} \cdot \underbrace{N_{sockets}}_{typically\ 2} \cdot N_{cores} \cdot \underbrace{N_{SMT}}_{typically\ 2}$$

logical cores to the user. Out of these, only the per-node cores are available via the shared memory programming techniques we discuss here, whereas only half of them are real physical cores with full hardware concurrency.

Modern supercomputers are hybrids of a distributed memory machine consisting of shared memory nodes. On this level, they fall into the category of MIMD. Within each core, they have hyperthreads (again a cooked-down MIMD flavour) which enable vector operations (SIMD). So SIMD and MIMD are combined with each other and MIMD comes along in multiple flavours (between nodes, between cores, between hyperthreads). Most core architectures finally support pipelining (MISD) giving us all three hardware parallelisation flavours of Flynn's taxonomy in one piece of metal.

Terminology

I use the term compute unit sometimes when I don't want to discriminate between core, node or thread. Unit is no formal term. I use it lacking a better word and to emphasise that we discuss generic concepts that are, too some degree, independent of the hardware realisation.

14.2   BSP

We next introduce one of the all-time classics on how to program a parallel (MIMD) machine: *Bulk synchronous processing* (BSP). BSP is among the oldest ideas of how to write a parallel program. It is a perfect example for a way to realise data parallelism on MIMD hardware.

In BSP, a program runs through four conceptual phases:

- It starts with some code which runs in serial. There's no parallelism here.
- The code then splits up into multiple parts (threads). Some call this a fork and speak of a fork-join model rather than of BSP.
- The threads now compute something in parallel. In the original BSP, there's no communication between the threads. The threads make up the bulk within the acronym BSP.

- Once all parallel activities have terminated, we fuse all the threads. This is the synchronisation step or join. It may comprise some data exchange as the forked threads are terminated.

There can be multiple of these fork-joins in a row.

To illustrate how BSP works in practice, we illustrate the program execution once more as graph for something we call a DAXPY. The acronym DAXPY means "double precision alpha x times y", i.e. $\alpha x + y$. α is a scalar while x and y are vectors. DAXPY and its single-precision counterpart SAXPY are omnipresent in scientific codes.

```
double r[N];
double x[N];
double y[N];
double alpha = ...;
for (int i=0; i<N; i++) {
 double alphaX = alpha * x[i];
 r[i] = alphaX + y[i];
}
```

The dependency graph (DAG) for this example is close to trivial and highlights that all the individual loop body executions can run in parallel. So a parallelisation of this code for two cores could conceptually look like

```
double r[N];
double x[N];
double y[N];
double alpha = ...;
for (int i=0; i<N/2; i++) {      for (int i=N/2; i<N; i++) {
 double alphaX=alpha*x[i];        double alphaX=alpha*x[i];
 r[i] = alphaX+y[i];              r[i] = alphaX+y[i];
}                                }
```

where the right code block is ran on one core and the left part on the other one. The initial setup of all data is done only on one core however. Such a parallelisation is exactly BSP (Fig. 14.1): There are a few initialisation steps where we create the arrays and initialise the α. These steps run on one thread only as there is no concurrency here. We then enter the bulk which we can split up into two or more chunks. This is

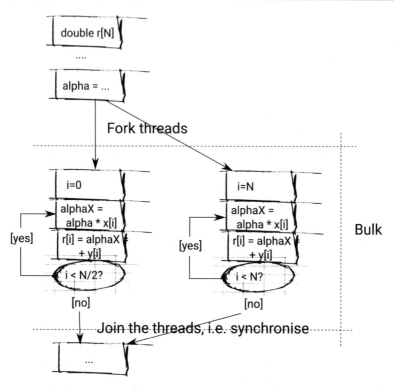

Fig. 14.1 Illustration of bulk synchronous parallel processing for a simple for loop realising a DAXPY. The code runs on two cores

where we *fork* the threads, i.e. now the threads are started. Each thread runs through a segment of the loop. There are *no dependencies* between the steps on one thread and the steps on the other thread. The steps are in no way synchronised or orchestrated.

Eventually, both threads have done their work and we join them. This is a synchronisation. We don't know how quick the threads are, whether the work is evenly split or whether they physically run on different hardware threads. Hence, we might have to wait for some to terminate. We finally continue serially.

Definition 14.3 (*Superstep*) In BSP, the individual bulk plus its synchronisation including the final information exchange forms a *superstep*.

14.3   Realisation of Our Approach

All of our programming techniques in this manuscript fall into one class: They are SPMD.

Definition 14.4 (*SPMD*) *Single program multiple data* (SPMD) means that all parallel compute units execute the same program (instruction stream).

This does not mean that they all follow the same instruction branch (if-statements), and it in particular does not mean that they run in locksteps. It simply means that they run, in principle, through the same piece of code. As such, the SIMD instructions we have seen—SIMD formally is a machine architecture model, i.e. has nothing to do with programming, so I write "SIMD instructions"—is a very special case of SPMD. We can even be more precise and read SIMD as very special case of BSP, where each instruction forms its own superstep and the join's synchronisation happens implicitly after the instruction. BSP is an example for SPMD programming, too. Every technique in this manuscript relies on SPMD. We never write two different programs that collaborate.

If you work on an application where you have a dedicated server program, a client software and a database software, then you have a classic example of a non-SPMD application. All three different components typically are shipped as different pieces of code.

Some interesting (historical) material

- L. G. Valiant: *A bridging model for parallel computation*, Communications of the ACM, Volume 33 Issue 8 (1990) https://doi.org/10.1145/79173.79181

> **Key Points and Lessons Learned**

- There are three classes of parallelism that we often find in computer codes: functional and data decomposition as well as competition. Data decomposition is the most popular approach.
- MIMD is the hardware architecture paradigm underlying multicore processors.
- In hardware specifications, we have to distinguish nodes, sockets, cores, and (hyper-)threads. As programmers, we typically handle (logical/software) threads which are mapped onto these machine concepts.
- BSP means bulk synchronous processing and describes a straightforward fork-join model where threads do not communicate with each other and synchronise only at the end. It is a paradigm how to design a data-parallel algorithm.
- BSP and traditional (SISD-focused) vectorisation can and are usually realised via SPMD programming.

Upscaling Models and Scaling Measurements

15

Abstract

Two different upscaling models dominate our work: weak and strong scaling. They are introduced as Amdahl's and Gustafson's law, and allow us to predict and understand how well a BSP (or other) parallelisation is expected to scale. We close the brief speedup discussion with some remarks how to present acquired data.

Let a simulation work on a big square. Assume that the square represents a huge image and the simulation computes something per pixel. The code runs for 24s on one core. This computation per pixel is where the 24s come from. How long will it run on two cores if we cut the square in two equally sized pieces? How long will it run on four cores if we cut the pieces once more?

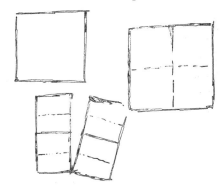

The natural answer is 12s and 6s. At the same time, we know that this is likely too optimistic. If we split our homework among our three room mates, they

likely have to chat, i.e. coordinate, with each other, and there might be tasks that
just can not be split among them or can not be split into equally sized parts.

Let another simulation run for 4s on four cores. Each core handles two pieces
of work. How much time would this run need on only one core? Finally we take
the setup and deploy it to 16 cores. At the same time, we also make it four times
bigger. For our image, we can for example increase the resolution in both the
x- and y-direction by a factor of two, so we get four times more pixels. What
happens if we run the simulation with four times as many compute resources
on a problem that is four times bigger?

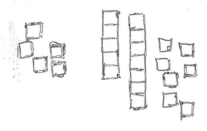

Before we start tuning or parallelising our codes, we have to ask ourselves what we
can and want to achieve. There are two reasons for this: On the one hand, we want
to get a feeling what we might end up with before we start the work. It is important
to assess whether parallelisation is worth the effort. On the other hand, we want to
assess the quality of a delivered solution: Have we delivered something reasonable?
For this, we first derive a model of our code:

Definition 15.1 (*Concurrency and concurrency level*) *Concurrency* means that
things can happen at the same time. The *concurrency level* is an upper bound on
how many computations could theoretically be run in parallel.

At any time of a simulation run, we may have a different concurrency level. However,
we can obviously compute an average one. In our BSP introductory example in Sect.
14.2, we have a concurrency level of N for the loop.

In the example, we then split up the loop into two parts only. There are many
potential reasons for this: The compiler or programmer might come to the conclusion
that the loop body is too cheap, and splitting it up into too small pieces induces too
much overhead. Alternatively, our computer might have only two cores, so there's
no point in splitting up the loop further, or we might want to map some concurrency
onto SIMD hardware features. In any case, we get a level of parallelism that is lower
than the concurrency level:

Definition 15.2 (*Parallelism*) *Parallelism* means that things are happening at the
same time.

In line with our previous discussion, one of the prime properties we are interested in is *scalability*. Scalability means how the runtime improves, i.e. goes down, when we add more and more compute units (cores or vector registers). It characterises the observed parallelism. Let $t(p)$ be the runtime of our code for p compute units. With this definition we can introduce two important quantities:

Definition 15.3 (*Speedup*) The *speedup* of a code is given by

$$S(p) = \frac{t(1)}{t(p)}.$$ (15.1)

I tend to forget what is in the numerator and what is in the denominator. But it is easy to reconstruct: If we have two processing units p instead of one, the time $t(2)$ should go down and we expect a speedup, i.e. $S(2) > 1$. So $t(1)$, the bigger one, has to be numerator.

Speedup tells us how fast we can make the code, but it does not tell us what price we have to pay for this, i.e. how much we have to invest:

Definition 15.4 (*Parallel efficiency*) The *parallel efficiency* of a code is given by

$$E(p) = \frac{S(p)}{p}.$$ (15.2)

Efficiency puts the scalability in relation to the number of compute units we invested.
There are some natural properties for both quantities:

- Both efficiency and speedup are positive. If not, we've computed something wrong.
- If the speedup falls below one, we've also done something wrong. $S(p) < 1$ means that we have invested more compute facilities, but our code has slowed down.
- In the best case, we have a speedup of $S(p) = p$. If we invest two compute units rather than one, we half the compute time. This is optimal. We call it *linear speedup*.
- The best-case efficiency therefore is one.
- If the speedup is $S(p) > p$, i.e. the efficiency $E(p) > 1$, we have to be sceptical. Sometimes, slightly superlinear speedup occurs due to caching effects for example. Yet, this is rare.
- The concurrency level yields a natural upper bound for the speedup.

15.1 Strong Scaling

Our first theoretical model for upscaling results from a simple speedup model. Introduced around the 1967s, Gene Amdahl started from the observation that many codes do some initialisation, then branch out and use many cores, but eventually have to synchronise all compute units again. As a result, we are never able to exploit all

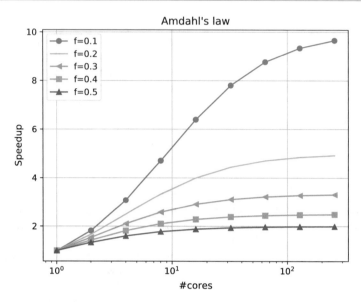

Fig. 15.1 Speedup curves for Amdahl's Law for various choices of f. The cores (x-axis) is the p from the formula, while the speedup is $S(p)$

compute units all the time. There is always some work which is (almost) sequential. The prime example for such codes is given by BSP.

Let $0 < f \leq 1$ be the fraction of code that runs serially. f close to 0 means that the code keeps all compute units busy almost all the time. This implies that its concurrency level is always equal or higher than the number of parallel compute nodes. f close to 1 means that the majority of our code is not running in parallel.

Definition 15.5 (*Amdahl's law/strong scaling*) The runtime of a code follows

$$t(p) = \underbrace{f \cdot t(1)}_{\text{serial work}} + \underbrace{(1 - f) \cdot \frac{t(1)}{p}}_{\text{parallel work}} .$$

The model is simple to explain: We start from a given time $t(1)$. A fraction $f \cdot t(1)$ of this time can't be parallelised. It does not benefit from more compute units p and thus "just is there" for a parallel run, too. The remaining fraction of the runtime $(1 - f) \cdot t(1)$ however is shared between the compute units, so it is reduced. The total runtime is the sum of these two ingredients.

Amdahl's law yields characteristic curves (Fig. 15.1): For small core counts, we see a significant speedup. The improvement due to additional cores becomes less significant with higher total core count. The code's parallel efficiency deteriorates. Eventually, the speedup plateaus almost completely.

Though Amdahl's law serves us well, it relies on some crude assumptions:

1. We can clearly separate serial from parallel code.
2. We neglect any kind of overhead effect. Everybody knows that coordinating more people requires more time. Yet, this model makes the runtime go down whenever we increase p.
3. We assume that the workload is something continuous that we can split up into smaller and smaller work packages which are then distributed. We know that real computer programs are not like that: There will be a point where we simply can't subdivide a problem further and run it on more compute units.

These shortcomings suggest that we often have to extend the law to reflect our particular cases. We have to augment it. It is however a good starting point to assess our code.

15.2 Weak Scaling

In our second theoretical model, we invert the train of thought. We assume that we have a piece of code that is running on a parallel machine. To make a statement on its speedup or efficiency, respectively, we ask ourselves "what happened if we ran this code without the parallelism". In 1988, John Gustafson argued along these lines. He still sticks to the same idea of a code, i.e. he still assumes that there is a serial part f in the code as in Amdahl's case.

Definition 15.6 (*Gustafson's law/weak scaling*) The runtime of a code behaves as

$$t(1) = \underbrace{f \cdot t(p)}_{\text{serial part}} + \underbrace{(1 - f) \cdot t(p)p}_{\text{parallel part}} .$$

This model is simple to explain, too: We start from a given time $t(p)$. A fraction $f \cdot t(p)$ is not running in parallel. When we rerun our code on one compute unit only, we have to do this part once. The remainder of the runtime is work that did run in parallel before. This whole workload now has to be processed by our single unit one by one. The total runtime is the sum of these two ingredients.

Gustafson's law gives us characteristic curves for speedup studies as we see them in Fig. 15.2: For small core counts, there is a price to pay to get the whole parallelisation thing going. Once we enter a regime with a reasonable number of cores, the code however performs well. The speedup curve "recovers".

Though Gustafson's law serves us well, it relies on some crude assumptions:

1. We can clearly separate serial from parallel code.
2. We neglect any kind of overhead effect. Everybody knows that coordinating more people requires more time, so there should be a penalty depending on p.

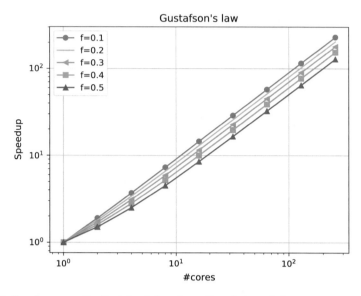

Fig. 15.2 Speedup curves for Gustafson's Law for various choices of f

3. The parallel code exhibits a certain f. The same f holds for a serial rerun. This
   implies that any parallel code we start from has a sufficiently high concurrency
   level.

These shortcomings suggest that we often have to extend the law to reflect our
particular cases. We have to augment it. It is however a good starting point to assess
our code.

15.3   Comparison

The two models behave very differently. One enters a saturation regime, the other one
suffers from some startup penalty before it recovers. To understand the difference,
we have to go back to our initial assumptions when we construct the formulae.

For Amdahl's law, we assume that we have a problem of size N and run this
problem on bigger and bigger machines. We can assume that N enters the time we
need to compute the result as a scaling factor subject to f. As it enters both $t(1)$
and $t(p)$, it is kicked out again immediately. The important thing to remember is:
We keep this N fixed as we scale up. Consequently, the f we observe for a single
compute node is the same for massively parallel runs. This is a *strong* notion of
upscaling.

Meaning of Strong Scaling

Strong scaling argues over the time-to-solution for a given problem setup.

For Gustafson's law, things are slightly different: We take a problem from the parallel machine and run this one on one unit. For both runs, we assume that the f is a constant and the same. To keep f constant in a real application, we might have to increase the problem size if we switch to bigger machines. Otherwise, parallelisation and synchronisation overhead will alter f. If we want to keep the ratio of serial code to parallel code constant, we have to make N bigger for bigger p. This observation at Sandia National Labs motivated John Gustafson to introduce the new law. The important thing to remember is: We increase the problem when we scale up to ensure that the ratio f remains constant. This is a *weak* notion of upscaling as we alter the underlying problem.

Meaning of Weak Scaling

Weak scaling argues over the throughput of an implementation, i.e. it keeps ratio of fully parallel to serial code parts invariant.

When we discuss the behaviour of our code, we have to make it crystal clear which type of upscaling model we use. We also have to be aware that either model might be ill-suited:

- If we work on a machine with limited memory and do not connect multiple of these machines, weak scaling does not arbitrarily make sense. At least, it is capped.
- If we work with an application where the accuracy depends on the problem size, strong scaling might make limited sense. Users will use a bigger computer to solve a bigger problem.
- If we work with a code that suffers from too much parallelism overhead, users will never scale up further. They will instead switch to other/bigger problems.

Depending on the model used, we obtain two different formulae for the speedup:

$$S_{\text{strong}}(p) = \frac{t(1)}{t(p)} = \frac{t(1)}{f \cdot t(1) + (1 - f) \cdot \frac{t(1)}{p}} = \frac{1}{f + (1 - f)/p}, \quad \text{v.s.}$$

$$S_{\text{weak}}(p) = \frac{t(1)}{t(p)} = \frac{f \cdot t(p)p + (1 - f) \cdot t(p)}{t(p)} = f \cdot p + (1 - f). \quad (15.3)$$

Algorithmic complexity

Whenever we compare timings for different problem sizes, we have to be very careful with the algorithm's complexity. If we measure runtimes for two different setups, we cannot directly compare them to say which one is better. Let our particle setup serve as an example: A straightforward implementation is in $\mathcal{O}(N^2)$, i.e. if we double the number of particles our runtime goes up by a factor of four. As a result, the $t(1)$s in the formula above are not the same and we cannot eliminate them directly by inserting one line into the other.

To fix this is relatively easy, once you know the complexity of your algorithm. A popular approach is not to report on runtimes, but to report on relative runtimes: If one setup it twice as big as the other and if our algorithm is in $\mathcal{O}(N^2)$, we divide the time for the bigger setup by four and then reconstruct f. In a particle context, we now report on "force computations per seconds" rather than runtimes. If we have a different complexity, we need a different scaling.

Determine f

Before you scale up a code, it is reasonable to determine f experimentally.

Determining f gives you a good indication how good your code should scale for large core counts. If you do not match your predictions for higher core counts, you subdivide the problem too rigorously and suffer from parallelisation administration overhead or you have a flaw in your parallel implementation.

15.4 Data Presentation

There are different ways to present speedup data. A convenient way is to plot the speedup, i.e. the quantity (15.1) over the processor/core count p. A typical plot is similar to

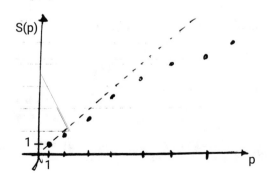

A dotted line represents the linear speedup. All data is calibrated towards the $(1, 1)$-point, as $S(1) = 1$ by definition. You have to decide carefully whether to connect your measurement points with a (trend-)line (cmp. Appendix F.2).

A typical plot flattens out or plateaus with increasing p. This saturation effect simply means that the problem starts to become non-decomposable and thus adding further processing units p does not yield any runtime improvement anymore. The stagnation in performance is a clear materialisation of strong scaling behaviour.

Two major issues with this plot are that

1. we do not see any absolute times, i.e. we cannot compare different trends;
2. we cannot plot any experiment where the experiment cannot be solved on one compute unit in reasonable time.

As an alternative to speedup plots, we can plot the efficiency (15.2). The fundamental problems remain the same.

If you want to give your reader an idea about real timings plus want to compare different setups where some setups are too big to fit onto a single compute unit, you can rely on some spider web graphs:

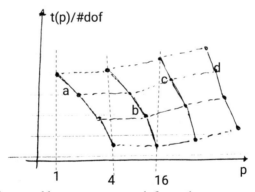

It is more difficult to read but presents more information.

- On the one hand, we do not present any speedup or efficiency, but we plot (normalised) timings for different compute unit counts.
- On the other hand, we present data for different experiments in one plot.

In the example above, we assume that experiment b is roughly four times as expensive as a, while c is four times more as expensive as b, and so forth. The calibration, aka what means expensive, can be tricky and it requires a careful description in the caption/text to make it clear to the reader what quantities are displayed. Once this is mastered, we directly visualise interesting properties: If we connect the measurements for one problem size choice (here labelled as (a), e.g.), we obtain a classic strong scaling curve. It should show a linearly decreasing trend. If we connect the measurements of different problem sizes which correlate—if (b) is four times the compute cost of (a), (c) four times of (b), …, we might connect the dots for $p = 1$, $p = 4$, $p = 16$—a flat line would indicate perfect weak scaling.

Often, scalability plots rely on a logarithmic scale. Ensure however that you pick a log scale for both the x-axis and the y-axis, as perfect scaling is called "linear scaling"; (cmp. wording above) and you want linear scaling to result in a line.

Some Interesting Material

- Michael McCool, James Reinders and Arch D. Robison, : *Structured Parallel Programming—Patterns for Efficient Computation.* Morgan Kaufman (2012)
- Gene M. Amdahl: *Validity of the single-processor approach to achieving large scale computing capabilities.* Proceedings of the American Federation of Information Processing Societies. AFIPS Press, pp. 483–485 (1967) https://doi.org/10.1145/1465482.1465560
- John L. Gustafson: *Reevaluating Amdahl's law.* Communications of the ACM (31), pp. 532–533 (1988) https://doi.org/10.1145/42411.42415
- David H. Bailey: *Twelve Ways to Fool the Masses When Giving Performance Results on Parallel Computers.* Supercomputing Review, pp. 54–55 (1991) http://crd-legacy.lbl.gov/~dhbailey/dhbpapers/twelve-ways.pdf

> **Key Points and Lessons Learned**

- We distinguish concurrency level (theoretic bound) and actual parallelism (observed).
- Speedup and efficiency are well-defined.
- Strong and weak scaling describe different scaling patterns and we have to be careful which one we choose.
- It is straightforward to derive the formulae for both upscaling models.

OpenMP Primer: BSP on Multicores

16

Abstract

We introduce OpenMP for multicore programming through the eyes of the BSP programming model. An overview over some terminology for the programming of multicore CPUs and some syntax discussions allow us to define well-suited loops for OpenMP BSP. Following a sketch of the data transfer throughout the termination of the bulk, we abandon the academic notion of BSP and introduce shared memory data exchange via critical sections and atomics. Remarks on variable visibility policies and some scheduling options close the chapter.

If you cook a proper meal following a recipe, then the recipe consists of steps. Your cookbook might contain paragraphs alike "fry A in batches and put aside, then add B and put everything in the oven; after that, take C and …". This is a sequential model which is optimised to make the best out of your time. If you have to do the same dish for your family's Christmas dinner (with enough space in your kitchen and lots of people who volunteer to help), your first step might be to rearrange these instructions such that you can all work in parallel.

If you write down the cooking instructions for this particular meal onto a fresh sheet of paper, you have two options: You can stick to the original recipe and add annotations who does what, i.e. which steps run in parallel. Alternatively, you can write down one recipe per helper.

The parallelisation of a code via directives typically consists of three steps:

1. Write down the code in such a way that it can be executed in parallel.
2. Make the code run in parallel by inserting the right directives.
3. Do some performance analysis and engineering.

Often, the three steps are part of an iterative process. The algorithm engineering of the first step is beyond the scope of this text. We focus primarily on the second

© The Author(s), under exclusive license to Springer Nature Switzerland AG 2021
T. Weinzierl, *Principles of Parallel Scientific Computing*, Undergraduate Topics in Computer Science, https://doi.org/10.1007/978-3-030-76194-3_16

step. However, any algorithm engineering, i.e. rewrite of code such that it exhibits parallelism, requires a sound understanding of what is possible on a machine, i.e. what is possible in step two.

For the "make it run in parallel" task, we rely on OpenMP. OpenMP originates from a multicore/BSP mindset. Although it has emancipated from it, the BSP heritage is still visible within the standard. Finally, new generations of OpenMP support GPGPUs besides classic multiprocessors.

Our strategy thus reads as follows: Since we are already familiar with OpenMP's vectorisation capabilities, we first introduce OpenMP within the classic multicore BSP context. The aim of our studies is to sketch the OpenMP mindset, to allow a quick start and to draw connections to more general computer science patterns. The goal is *not* to provide a comprehensive OpenMP introduction or description—there's enough (better) material available. Starting from the introductory BSP, we discuss more advanced concepts—mainly OpenMP's tasking—which not only generalise and eventually emancipate from BSP but also allow us to realise functional decomposition. Finally, we return to BSP parallelisation to sketch how this fits to modern GPGPUs.

16.1   A First (Working) OpenMP Code

The core compute routines in Chap. 3 are the force calculations and the particle update. We can straightforwardly parallelise them with BSP:

```
#pragma omp parallel for
for (int i=0; i<N; i++) {  // for every body in system
  // reset forces
  force[i][0] = 0.0; force[i][1] = 0.0; force[i][2] = 0.0;
  for (int j=0; j<N; j++) { // compute interactions
    if (i!=j) {                  // but only with other objects
      double Gx, Gy, Gz;
      calcForce( p[i][0], p[i][1], p[i][2],
        p[j][0], p[j][1], p[j][2], Gx, Gy, Gz
      );
      force[i][0] += Gx;         // accumulate forces
      force[i][1] += Gy; force[i][2] += Gz;
    }
  }                              // forces are there, now we can
}                                // update particles (don't mix)

#pragma omp parallel for
for (int i=0; i<N; i++) {  // for every body in system
```

```
    p[i][0] += dt * v[i][0]; p[i][1] += dt * v[i][1];
    p[i][2] += dt * v[i][2];
    v[i][0] += dt * F[i][0]/m[i];
    v[i][1] += dt * F[i][1]/m[i];
    v[i][2] += dt * F[i][2]/m[i];
  }
```

This is a straightforward BSP parallelisation, and I illustrate its behaviour for 12 objects and three cores in Fig. 16.1. The key observations read as follows:

1. There are two supersteps. These are the two outer for-loops.
2. Whenever the code hits a superstep, it spawns three OpenMP threads.
3. OpenMP automatically splits up the for-range into three chunks.
4. Once a thread has done its part of the iteration range, it waits for the others to join again. This synchronisation happens implicitly as we hit the end of the for-loop.

Run multithreaded code. Before we dive into details, we discuss how to run OpenMP multicore code.

An executable translated with OpenMP should automatically link to all required libraries. The software that's integrated into your code by the compiler to take care of all of your parallel needs is called the *runtime*. This runtime has to know how many threads it should use on your system. For this, it evaluates the environment variable OMP_NUM_THREADS. On bash, type in

```
    export OMP_NUM_THREADS=3
```

before you start your code. It makes OpenMP use three threads. By default, most systems set OMP_NUM_THREADS to the number of logical cores; this includes hyperthreads.

Thread team (fork). Whenever an OpenMP code hits a #pragma omp parallel statement, it interprets the block following the pragma as superstep

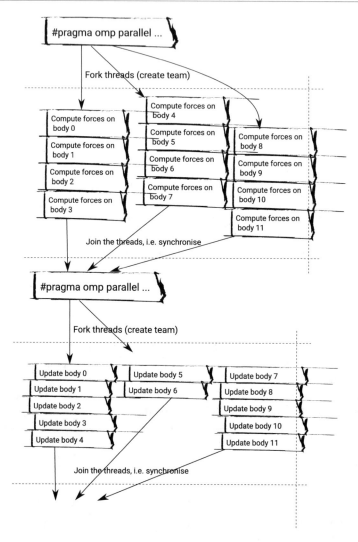

Fig. 16.1 Schematic illustration of an example execution of the first simple OpenMP parallelisation

in a BSP sense. Upon entering the pragma's block which is the loop following it in our example, OpenMP fires up OMP_NUM_THREADS threads. The thread count is set at runtime; which is great for scalability studies but also makes your code platform-independent. If you run it on a different node with another core count no recompile is required. At the end of the block, all threads of a team join again.

Definition 16.1 (*Thread team*) Whenever OpenMP hits a parallel region, it spawns up one *thread team* with OMP_NUM_THREADS threads.

While modern OpenMP is very good at administering the teams it might make sense to distinguish the team creation from the actual work distribution. Historically, the team's threads were literally destroyed at the end of the region and re-created later on. This induces overhead. Today, OpenMP implementations often let threads idle and reuse them, i.e. wake them up when you hit the parallel region.

```
#pragma omp parallel
{
 #pragma omp for
 for (int i=0; i<N; i++) {  // for every body in system
  [...]
 }                          // update particles (don't mix)

 #pragma omp for
 for (int i=0; i<N; i++) {  // for every body in system
  [...]
 }
}
```

Loop split. `#pragma omp parallel for` is a shortcut for `#pragma omp parallel` (to spawn/fork the threads) followed by `#pragma omp for`. The latter is OpenMP's pragma for for-loops. It ensures that the loop range is properly split up and the loop iterations are distributed among the threads of the team. Such combinations of a for and a parallel directive are called *combined constructs* in OpenMP.

Parallel region without a range split

If you forget the `for`, you will end up with `OMP_NUM_THREADS` threads which all execute the whole for-loop. This is obviously a bug—you don't want the loop to be run `OMP_NUM_THREADS` times in parallel.

Synchronisation. At the end of the region following a `#pragma omp parallel`, OpenMP automatically joins all threads.

Definition 16.2 (*Barrier*) A *barrier* in OpenMP is introduced via the pragma

```
#pragma omp barrier
```

All threads of the team have to enter this point before they continue. The barrier thus is a *collective* operation.

The end of a parallel region implicitly introduces a barrier, i.e. you can add a barrier statement or leave it out—OpenMP will always add one. This barrier precedes the actual join.

While barriers are rarely used in OpenMP applications, their existence (and the discussion of their existence) emphasises that OpenMP has been specified with a MIMD architecture in mind. That is, the individual threads run asynchronously, and you never know which thread processes which statement at a certain point. In particular, the threads are not in lockstep mode. OpenMP's simd statement is explicitly the opposite thing: Here, all steps run in lockstep. Indeed, we can (logically) interpret a simd region as a sequence of instructions where each statement is followed by a barrier.

OpenMP advocates a pragma-based parallelisation. However, it defines a few library functions as well. I never use them, though sometimes they are good for debugging, e.g.

- omp_get_max_threads returns the number of threads you have in your code. This should be the same number that you did set via the environment variable OMP_NUM_THREADS.
- omp_get_num_threads gives you the number of threads that are actually used within a parallel environments. They might be less than the maximum number of threads.
- With omp_set_num_threads you can set the number of threads manually. This does not always work. If your operating system constrains the number of threads (most supercomputers do), then I found my code often able to reduce the number of threads, but it was not able to exceed OMP_NUM_THREADS.
- omp_get_thread_num returns the current thread number.

OpenMP functions

Try to avoid OpenMP's functions. Work with the pragmas only.

16.2 Suitable For-Loops

To identify loops that are well-suited for OpenMP parallelism, I recommend to sketch the control flow via a graph first. OpenMP regions become apparent by a graph that fans out and fans in again.

Loops that fit to a multicore computer

In line with Sect. 7.2, not every loop is well-suited for OpenMP parallelism either. They have to be in *canonical form*:

Definition 16.3 (*Canonical form of OpenMP loops*)

1. *Countable.* We have to know how often we run through the loop when we hit it for the first time. This allows OpenMP to distribute a loop over multiple threads. Something like a while loop where we don't know how often we run through it is ill-suited for BSP parallelism.
2. *Well-formed.* Analogous to SIMD loops, a canonical for-loop once more has to have a simplistic counter increment/decrement and the ranges may not change at runtime (see countability requirement). You may not jump out of a loop and you should not throw an exception, as OpenMP in general does not deal with exceptions.

On top of the "hard" constraints formalised by the term canonical form, loops have to exhibit a reasonable *cost*. Launching a parallel loop is expensive. Different to compiler-based vectorisation, OpenMP is prescriptive. Analogous to its `simd` directive, an annotated loop will definitely be parallelised. It is thus important that you are sure that your loop chunk is reasonably expensive.

Compared to the SIMD requirements, there are delicate differences of a canonical loop for multithreading compared to a loop that can be vectorised: Both flavours of canonical require the loop count and range to be fixed and simplistic such that OpenMP can straightforwardly split it up. It is particularly important that there are no dependencies between the loop body and the loop count. You may not alter the loop counter within the loop.

Different to the vectorisation, "canonical" for multithreading is less restrictive for the loop body. Complex branching and function calls are allowed. This is not a surprise given that we execute the loop on different threads, i.e. full logical von Neumann cores. Each thread has its own thread count and speed, and can arbitrarily jump through the source code. This is a direct consequence of a programming paradigm made for MIMD machines. There's a further small difference: Parallel regions can be nested.

Definition 16.4 (*Flavours of nested*) If an OpenMP block is used within another OpenMP block, it is *nested*. A parallel OpenMP block is *strictly nested* within another parallel block, if there is no code between the outer and the inner block. It is *closely nested*, if there is no other parallel region nested between them: If a SIMDised loop A is embedded within a parallel loop B within a parallel loop C, A is closely nested within B, but not within C. Finally, we formally distinguish *worksharing* regions from regions that are not worksharing. A worksharing region is one that makes OpenMP run different instructions on different threads. An *omp parallel* alone is not worksharing, as it only fires up threads. A *for* however splits up work.

SIMD loops can be nested within parallel regions, but not the other way round. If you use a nested parallel loop within a nested parallel loop, this works, but you pay for in terms of adinistration (scheduling) overhead; though every new OpenMP generation becomes better in handling this from a performance point of view.

Both SIMD and MIMD parallelisation rely on source code with sufficient cost, as the parallelisation comes with overhead. The notion of cost however differs. OpenMP parallelisation suffers from a thread start-up cost. After the thread kick off, the underlying loop range is split up and distributed over the threads. As a consequence, either the loop range has to be vast—individual threads then are assigned a significant chunk of the iteration space—or the individual loop iteration has to be expensive. In both cases, the cost per thread is high and the parallelisation overhead is amortised.

In a SIMD environment, thread divergence reduces the effective average cost of a loop—if only half of the registers are really used, we can read this as "half of the compute cost per instruction". It renders a loop less likely to benefit from SIMDsation. `parallel for` statements are not sensitive to complex if statements as long as the cost per loop iteration remains high or each thread is given a big enough range to run through. Since the MIMD parallelism does not rely on lockstepping, thread divergence is not an issue and techniques like masking or blending do not play a role. Yet, thread ill-balancing—if one thread needs more time than the others—eventually harms the efficiency.

simd vs. parallel for

Within a cascade of loops, `simd` has to be applied to the innermost loop. `parallel for` should be applied to a rather outer loop, as the workload per loop iteration should be high, and we have to avoid synchronisation.

16.3   Thread Information Exchange

In the pure, academic BSP model, threads do not communicate while they compute. They run totally independent of each other. Only at the end, they synchronise through an (implicit) barrier, i.e. they wait for each other. This is where they also combine their individual results. We make threads work on their local (partial) result (data decomposition/parallelism paradigm) before we bring these partial results together. The latter is an all-to-one data flow. It is a *reduction* (cmp. Definition 16.4).

OpenMP as tool to realise parallel codes does not restrict itself to collective, BSP-style reductions only: We can use the shared memory as a scratchpad to which some threads write data while other threads pick it from there. For both data flow types—the all-to-one and the anarchic paradigm where everybody scrambles into the memory as a big scratchpad—-we need some synchronisation mechanisms. Otherwise, our data quickly become inconsistent:

Definition 16.5 (*Race condition*) A race condition is a situation where two threads access a variable in a non-deterministic fashion without proper orchestration. One of them hence "wins". They "race" for superiority. We distinguish three types of *race conditions*:

1. *Read-write race condition.* Thread A writes to a location and thread B reads from this location. If thread B requires data from A through a shared variable yet A is too slow (or B too quick), then B reads the variable before A has actually dumped the result.
2. *Write-read race condition.* Thread A writes to a location after thread B reads from this location. If thread A is quicker than B, then A has overwritten the variable both have access to before B fetches the result, i.e. A feeds B with new data though B has been interested in the old value.
3. *Write-write race condition.* Both A and B write to a variable. Whoever is the slower thread overwrites the other result and is therefore the "winner".

As OpenMP is constructed around a shared memory mindset, i.e. all data are by default visible to all threads, the example below produces data races:

```
#pragma omp parallel for
for (int i=0; i<20; i++) {
  r += x[i] * y[i];
}
```

The code determines an inner scalar product (stored in r). Let's assume this code is executed on two cores and OpenMP splits up the loop range into 0–9 and 10–19. There are multiple data races (Fig. 16.2): If both cores add something to r, they first have to bring r into their register. They then add $x[i]*y[i]$ to the register content

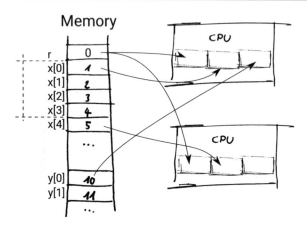

Fig. 16.2 Illustration of a parallel scalar product evaluation with data races. If we assume that the arrays x and y from $r = x \cdot y$, $x, y \in \mathbb{R}^8$ are split into two times four on a dual core machine, then both cores load the value of r from the memory as well as x[0] and y[0] or x[4] and y[4], respectively. They multiply the x and y component. After that, they have to write back r at one point. As we lack any synchronisation, there is a risk that one core overwrites the result from the other one

and write the result back to memory. If thread A and B write concurrently to the memory location and if B writes late, then the contribution of A will be overwritten and consequently lost. We have a classic write-write where we don't know whether A's or B's result "will rule". Furthermore, there's a write-read (B should have taken the result from A).

Finding data races

Data races are nasty bugs as they are often very hard to find. However, there are tools out there which can search for data races in your code. Use these tools when you search for bugs!

16.3.1   Read and Write Access Protection

Definition 16.6 (*Critical section*) A *critical section* is a named sequence of instructions. While one thread executes instructions from the sequence, no other thread is allowed to "enter" the section. That is, the very moment one thread starts to process the first instruction of the critical section, it "locks" the sequence. No other thread can start to process this section as well. Other threads have to wait until the first thread has done all the instructions within the critical section.

```
#pragma omp critical (mysectionA)
{ x *= 2 ; }

#pragma omp critical (mysectionB)
{ y *= 2 ; }

#pragma omp critical (mysectionA)
{ x /= 2; }
```

If we use the above code snippet within a parallel environment of OpenMP, no two threads can concurrently modify x. OpenMP does not enforce us to name a section (in round brackets), but any unnamed critical section blocks all other (unnamed) sections globally. Two critical sections can be in totally different branches and still exclude each other. Nested critical sections allow us to realise complex exclusion logic such as "A can run if B doesn't run, but B may not run concurrently to C". They also allow us to produce deadlocks.

Definition 16.7 (*Deadlock*) A *deadlock* arises if a thread A waits to be admitted to enter a critical section x while it already is in a critical section y. At the same time, thread B has already entered a critical section named x and waits for a access to y before it can leave x.

Deadlock means that your program does not progress anymore (Fig. 16.3). A deadlock often is kind of the counterpart of a data race. Where a data race results from a lack of data access protection, a deadlock results from too aggressive protection. The easiest way to find a deadlock is to use a debugger, to kill the execution once it hangs, and then to study the backtrace. Sporadically occurring, non-deterministic deadlocks however remain nasty.

The concept of different critical sections in different code parts works, once again, due to the fact that OpenMP does not imply lockstepping. The individual threads run through "their" code independently. Indeed, a real critical section is difficult to realise in a lockstepped environment, as critical sections can protect a whole block. You can mask out all other threads besides the main thread, but there is no way to stop some lockstepped threads from progressing through the code; non-trivial workarounds are required:

Critical sections with lockstepping

Critical sections can be realised in a lockstepping environment by looping over all threads with a simple for loop. Per loop iteration, only one thread is allowed to "do something" (aka "run through the critical section"). Such a pattern *serialises* the run-through through the critical section manually.

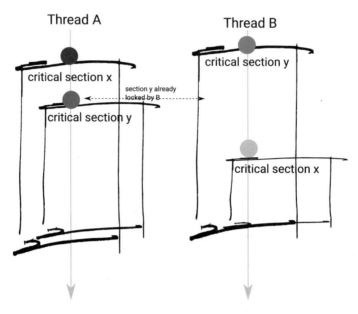

Fig. 16.3 Classic deadlock setup: A thread locks one critical section (dark red), and starts to wait for permission to enter a second one (lighter red). This second section however has been entered by another thread before (dark green) which in return waits for permission to enter A's section (light red)

The code snippets above use brackets even though they protect only a single statement each. The brackets are syntactic sugar. The code semantics are preserved without them. Unfortunately, critical sections are slow. For faster "critical access", i.e. race-free access, to a single built-in variable, we switch to atomics:

Definition 16.8 (*Atomics*) Critical sections are expensive. Administering them induces severe overhead. If one individual variable access is to be protected, *atomics* are a better option. Atomics work only for primitive types such as integers.

```
#pragma omp atomic
x++;
```

Semantically, atomics are not different from critical sections around a single statement. However, they are significantly faster as they are natively supported by the hardware. Chips have special "wires" for atomics.

As atomics apply only to a single statement, there is no way to label them. Actually, the system uses the variables' locations in memory as label, i.e. to identify two atomics which exclude each other. Atomics wrap around one single variable access. You can read from a variable

```
#pragma omp atomic read
y = x;
```

in which case the read from x is protected, you can write to a variable

```
#pragma omp atomic write
y = x;
```

in which case the write to y is protected, or you can update:

```
#pragma omp atomic update
x++;
```

In the latter case, we have a read-write, i.e. the variable is read, incremented and written back; all is protected. The default clause of an atomic, i.e. the one used if you specify none, is update.

Atomic updates with a second variable

If you use atomics to update a variable (increment, e.g.), the right-hand side should not be a complex expression. If you protect x+=foo()-y, the evaluation of foo()-y is not thread-safe.

There's a fourth useful nuance of an atomic: the capture. It is a natural extension of update which updates the variable plus returns the variable content prior to the update. A capture is the only atomic that accepts two statements: the backup of the value into another value plus a follow-up modification:

```
#pragma omp atomic capture
{
  oldValue = x; x++;
}
```

Although written down as a two-liner, this is effectively a single atomic statement.
With the update statement, it is possible to write the *compare-and-swap (CAS)*:
We wrap oldValue = x; x=y; into a captured region and follow-up with
oldValue==comparison.

With atomics and critical sections, we can discuss different correct versions of
the reduction within the scalar product.

```
#pragma omp parallel for
for (int i=0; i<20; i++) {
  double tmp = x[i] * y[i];
  #pragma omp atomic
  r += tmp;
}
```

is a valid implementation of the reduction, as it protects the access to the result r with
an atomic. It is not fast however. Even though atomics themselves are not expensive,
this code effectively serialises the scalar product computation. Only the evaluation of
the temporary results in tmp exhibits some concurrency. If we replaced the atomic
with critical, we would get the same result, but the code would be even slower.
The variant

```
#pragma omp parallel
{
  double myR = 0.0;
  #pragma omp for
  for (int i=0; i<20; i++) {
    myR += x[i] * y[i];
  }
  #pragma omp atomic
  r += myR;
}
```

is significantly faster. Here, we first fork our threads and create a new helper variable myR. The following OpenMP `for` then splits up the loop range, and we accumulate the thread-local result over a chunk of x and y within our local helper. Once an individual loop has terminated—this part runs totally in parallel on the individual threads—we add the local accumulation result to r. This last step is protected by an atomic.

The sketched data flow from Fig. 16.4 is an all-time classic of parallel computing. It is nevertheless ugly, as it requires us to alter the original code significantly. This does not commit to an OpenMP philosophy where parallelisation should be an add-on and code should remain unchanged and working if the compiler does not understand OpenMP. The annotation standard indeed provides an alternative which is as efficient yet a one-liner: reduction clauses. Before we discuss this particular type of all-to-one communication (all threads accumulate their partial results into one variable), we however first discuss variable visibility; a topic we have implicitly taken for granted so far.

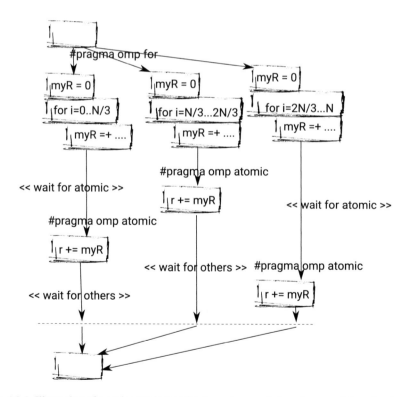

Fig. 16.4 Illustration of a scalar product realisation where each thread accumulates a local result in myR. This local result finally is added to the global outcome. The roll-over is protected by an atomic

16.3.2 Variable Visibility

The discussion so far has assumed that all variables within our loops are shared if they are declared outside of the loop body. This is due to our sloppy remark that we work within a shared memory environment. Upon close inspection, the assumption that all data not declared within the loop body are shared is already wrong: If we parallelise a for-loop with OpenMP, each thread gets its own iteration range. That means, if we run a statement $for(int\ i = 0; i < N; i + +)$, each thread has its own notion of (a modified) N plus initial value for the loop counter, and, in particular, its own i. If all threads shared all variables, all would use the same i. As the OpenMP code yields the correct result, we conclude that OpenMP has, without our interference, not shared these variables naively. Indeed, OpenMP distinguishes different variable visibilities.

Definition 16.9 (*Shared variables*) By default, all variables that have been declared outside of the OpenMP region are *shared*. Shared means that all threads have access to the same memory location. Loop counters and loop range thresholds are excluded from this convention.

There is no need to tell OpenMP explicitly that variables decleared outside of the OpenMP region are shared, though you can do so if you want:

```
double r = 0;
#pragma omp parallel for shared(r)
for(int i=0; i<size; i++) {
  r += x[i] * y[i];
}
```

Private variables are variables that are owned by the respective thread. They are not visible to other threads. We have two options to create private variables: If we declare a variable within the parallel region, it is a local variable and hence private, i.e. each thread creates it on its own.

```
#pragma omp parallel
{
  int myPrivateVariable;
}
```

ensures that each thread has its own working copy of `myPrivateVariable`. We use this technique when we define `myR` for the reduction (Fig. 16.4). Though this is a natural way to create thread-local variables in C, introducing helper variables in the original code just to ensure that it is not broken by a pragma is not 100% in line with our OpenMP philosophy. We want to stick as close to a serial code as possible and add all parallelisation through pragmas only.

Indeed, OpenMP allows us to specify private variables explicitly within pragmas. This means that the variable labelled as private is not used once OpenMP kicks in. Instead, OpenMP ensures that one instance of the specified variable is created per thread. In the example

```
double r = 0;
#pragma omp parallel for private(r)
for(int i=0; i<size; i++) {
  r += x[i] * y[i];
}
```

nothing happens to the actual variable `r` declared in the first line. OpenMP hits the fork instruction, splits up the loop range, and creates a new instance of `r` per thread. Each thread now works against its own instance. Once the threads join again (implicitly at the end of the for-loop), all these local result copies are destroyed and lost. In the above example, the private clause breaks the code semantics.

Definition 16.10 (*Private variables*) *Private* variables are thread-local variables, i.e. there's one instance per thread.

To understand the implications of this definition, we keep in mind that there are no default values in C: Whenever OpenMP creates a thread-local variable, it is not properly initialised. It holds garbage. If we want a private variable to be initialised, we have to do this ourselves within the OpenMP parallel block `r=0`. Yet, OpenMP allows us to declare a variable as `firstprivate(r)`. In this case, each thread-local variable is initialised with a copy of the original variable. There is also a lastprivate. I've never used the latter.

Visibility modifiers

Prefer private and firstprivate over an explicit creation of thread-local variables. They help to keep your code the same and, hence, valid even if no OpenMP is available.

Loop counters are by default made private, and it is the OpenMP runtime which initialises them properly. The same holds for the lower and upper loop bounds. It is the runtime that decides how to initialise the variables. It knows at runtime how big the actual loop count is and how many threads are to be used.

Parallel for

Whenever you write

```
#pragma omp parallel for
for (int i=iMin; i<iMax; i++) {
```

OpenMP implicitly makes

- iMin and iMax private. Once the thread-local copies are created, it initialises the values appropriately. Logically, the values are something closer to firstprivate, as OpenMP automatically sets them to the right value. In this particular case, this "right" value is not a plain copy of the iMin's or iMax's original value, but one that provides meaningful boundaries;
- the counter i private. The C++ expression automatically sets it to the right iMin per thread before we run through the loop body for the first time.

16.3.3   All-to-One Thread Synchronisation

So far, all of our OpenMP threads run asynchronously. Once the fork is done, each thread runs through its source code with its data. At the end of the OpenMP region, the threads all join again. They synchronise with each other. OpenMP's synchronisation at the end of a BSP superstep is implicit, i.e. nothing is to be done on the programmer's side.

A barrier is the simplest collective we can think of. It synchronises threads yet does not allow them to exchange data. If threads exchange data towards the end of their lifetime, we perform a reduction. Reductions are an advanced type of a collective, which we can either implement manually or realise through built-in mechanisms. They are more specific than a barrier as they synchronise, plus they move information around, plus they compute something:

Addition, multiplication or the minimum or maximum function are frequently used binary operators within reductions. However, we can also use binary/logical ors and ands. While we can insert barriers literally everywhere, OpenMP's reductions close the BSP superstep. You cannot insert them manually into a specific place in your loop your code.

With the reduction, we can finally write our scalar product routine efficiently:

```
#pragma omp parallel for reduce(+:r)
for(int i=0; i<N; i++) {
  r += x[i] * y[i];
}
```

The reduce operation is given an associative and commutative operator (addition). As we specify which variable is to be used to reduce the datum, the compiler automatically declares this variable as private, i.e. each thread hosts its own copy of result. The copy furthermore is initialised with zero, all accumulations are thread-local, and only the result is reduced into the original, outer result variable at the end. The zero stems from the following convention:

All of OpenMP's built-in reduction operators have a *neutral element*. For the addition, 0 is the neutral element as you can always add 0 to a datum. It doesn't change the value. For the multiplication, 1 is the neutral element. If a programmer uses an OpenMP reduction, the reduced thread-local variables are automatically initialised with the neutral element.

The reduction keyword yields significantly faster code than a manual solution, as we outsource its efficient realisation to OpenMP (which typically knows quite a lot about the hardware and efficiency patterns). It might, for example, automatically construct a parallel-serial reduction pattern as we have sketched it before with the local myR, or it might decide to realise the combination of the partial results via a binary tree.

With OpenMP's reduction keyword, we can combine various reductions. It is possible to pass + for one operator yielding a sum ($\sum_i$) over all threads, while a second variable computes the maximum. The individual elements are separated by a comma.

Nondeterministic output

If you deploy the reduction's realisation to the OpenMP runtime, you have to accept that the output is nondeterministic. We know that the order in which we perform a reduction affects the outcome in floating point precision. With OpenMP's reduction, we have no influence on the order, and the order might change as users set OMP_NUM_THREADS to different values. For most of the codes I've worked with, this is a negligible effect. There are however application areas, where you should, in theory, keep this nondeterminism in mind.

16.4 Grain Sizes and Scheduling

OpenMP's parallel for leaves it to the OpenMP runtime to decide how to split up the iteration range. This way, one OpenMP code runs for different core configurations—we only have to change the environment variable OMP_NUM_ THREADS prior to the runs. Internally, OpenMP relies on the simplest heuristic possible to split up a loop: static scheduling.

Definition 16.11 (*Scheduling*) *Task scheduling* or *work scheduling* describes the decision (algorithm) which work is associated/deployed to which thread.

Scheduling algorithms for iteration ranges work internally with grain sizes.

Definition 16.12 (*Grain size*) The *grain size* of a decomposition is the smallest number of parallel work items (loop iterations) that are grouped into one work agglomerate for a thread.

The grain size used by a scheduler determines the granularity of the work decomposition. In theory, a loop with N items could be decomposed into N work packages that all run in parallel (Fig. 16.5). The grain size then would be one. However, any decomposition comes along with an overhead: Work packages have to be managed, synchronised, tracked. Therefore, OpenMP usually uses a grain size that is bigger:

Definition 16.13 (*Static scheduling*) OpenMP's default scheduling is *static*. A workset with N items is cut into pieces (grains) of size N/p ($N > p$) or one ($N \leq p$), respectively, and these pieces are then distributed among the p compute units. This

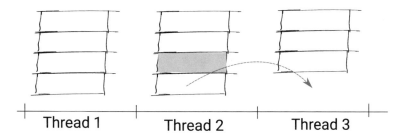

Fig. 16.5 Schematic illustration of static scheduling over a loop with 11 entries with a grain size of one over three threads. The work is reasonably balanced (4 vs. 4 vs. 3) over the work queues of the three threads. If the seventh entry (assigned to thread 2; marked) however needs significantly more compute time than the other work packages, we run into ill-balances. With dynamic scheduling, the thread three in such a case might steal a work package originally assigned to thread two (red dotted arrow)

happens when the code hits a loop. The distribution is not changed anymore at runtime.

The grain size of the partitioning hence is $\max\{N/p, 1\}$. Most implementations follow the following pattern: Each thread (spawned by `parallel`) has a queue of work items. When the code hits a `parallel for`, it creates work packages of grain size. They are continuous segments of the loop iteration range. Next, it distributes these over the thread queues. The threads take their workpackage from their queue and process their iteration range.

Users can alter the grain size by adding

```
#pragma omp parallel for schedule(static,20)
```

to their loops. This constrains the scheduling and ensures that no chunk of the iteration range is smaller or significantly bigger than 20 entries. We explicitly specify a value for the grain size.

This implies that either more than p chunks are distributed among the p work items—which is an artificial overdecomposition which in turn can lead to ill-balance—or it means that we use a smaller number of threads than what is actually available.

Interleaving work items

Most OpenMP implementations distribute the chunks in a round-robin fashion, i.e. core 0 gets the first chunk, core 1 the second, and so forth. With the default static decomposition, this means that the first core gets array items $[0, N/p[$, the second one $[N/p, 2N/p[$, and so forth. If you manually set the grain size to one, most runtimes will give you an interleaved task decomposition, i.e. the first core will get the array elements $0, p - 1, 2p - 1, \ldots$

Another application for a manual grain size is that you might find out that parallelisation pays off if and only if the underlying for loop is sufficiently big. In this context, OpenMP's if is worth mentioning:

```
#pragma omp parallel for if(a>20)
```

It switches OpenMP dynamically on or off. You can put in almost any boolean expression into the if; in particular the array size. As an alternative to the if condition or the grain size, you can also specify the exact number of threads used by a loop. I do not recommend the latter—a hard-coded thread count prohibits users to modify/tune the parallel environment via the command line and makes your code less platform-independent.

Grain sizes and if guards

Determine experimentally on two cores if parallelisation of a certain loop pays off or not. If it pays off only for reasonably big workloads, start to think about minimal grain sizes or an if guard such that the OpenMP parallelisation kicks in only if the workload is sufficiently big.

Definition 16.14 (*Dynamic scheduling*) *Dynamic scheduling* means that OpenMP threads are allowed to steal work from other threads if they idle.

The grain size for static partitioning helps us to eliminate unreasonable small work units per thread. Another problem arises if some loop iterations last significantly longer than others. Assume the first ten iterations of a loop with 20 elements last ten times longer than any of the remaining ten iterations. In this case, static partitioning with two threads will yield a workload distribution of 10 vs. 100, i.e. an extremely ill-balanced one. To eliminate this problem you can use dynamic scheduling

```
#pragma omp parallel for schedule(dynamic,2)
```

where the problem is cut into chunks of 2 (feel free to use another constant) and equally distributed among the work queues. So far, the scheduling does not differ from static scheduling. However, as soon as one queue is empty, the corresponding thread is allowed to steal a work package from another queue. The code starts with some initial estimate what it thinks might be well-balanced and dynamically balances work packages after that. Obviously, this dynamic balancing is not for free.

Dynamic scheduling

Use dynamic scheduling if you are sure that the workload per loop iteration fluctuates significantly and if the individual work packages are reasonably expensive.

We summarise that we have two types of scheduling flaws: Imbalances resulting from inhomogeneous workloads have to be tackled via dynamic scheduling. Inefficiencies due to too small work units have to be tackled via appropriate guarding mechanisms, i.e. minimal grain sizes, ifs, or explicit restrictions on the number of threads.

OpenMP provides a third scheduling flavour called guided. It is dynamic yet works with exponentially decreasing subproblem sizes. The ones distributed initially are large, the next ones only half of the size and so forth. This way, it tries to bring together the best of two worlds: the large grains scheduled initially ensure that not too much overhead is caused by over-decomposition. The small grains scheduled finally ensure that some small work packages remain available that can be used to balance out different threads.

Future OpenMP generations will likely offer further bespoke scheduling variants and tailoring opportunities.

Some important material

- J. LaGrone, A. Aribuki, C. Addison and B.M. Chapman: *A Runtime Implementation of OpenMP Tasks*. B.M. Chapman, W.D. Gropp, K. Kumaran and M.S. Müller: OpenMP in the Petascale Era. LNCS vol 6665, pp. 165–178 (2011) https://doi.org/10.1007/978-3-642-21487-5_13

Key points and lessons learned

- OpenMP realises the fork-join model through the `omp parallel` pragma. The behaviour of this pragma is controlled via the environment variable `OMP_NUM_THREADS`.
- Loops are automatically split and OpenMP takes the thread count and loop range into account. With the scheduling and grain size information as well as the if, we can control how this splitting behaves.
- Synchronisation and orchestration between threads is coordinated via collectives and critical sections plus shared memory and atomics.
- Reductions can be tailored through clauses and automatically pick the right neutral element to initialise the thread-local reduction variable.
- Important visibility modes of OpenMP are `private`, `shared` and `firstprivate`.

Shared Memory Tasking

17

Abstract

Starting from a brief revision of task terminology, we introduce task-based programming in OpenMP and the required technical terms. After a summary of variable visibilities for tasks—all concepts are well-known from the BSP world though some defaults change—we conclude the task discussion with the introduction of explicit task dependencies.

You are given the following code from a bigger project. Your colleagues ask you to explain what the programmer, who since left the company, wanted to achieve:

```
#pragma omp parallel for schedule(static:1)
for (int i=0; i<2; i++) {
 if (i==0) {
  foo();  // function defined somewhere else
 }
 else {
  bar();  // yet another function
 }
}
```

Which parallelisation paradigm do we see here?

Although OpenMP originates from a data decomposition/BSP world, it has emancipated from this mindset. On the one hand, scientists want to model more things than only plain data parallelism with a fork-join model. On the other hand, the explosion of parallelism on the hardware side implies that a sole BSP approach underperforms. No matter how good we parallelise, we quickly run into the situation where the serial part (in an Amdahl sense) between BSP supersteps dominates the code's efficiency.

17.1 Task Graphs Revisited

Definition 17.1 (*Task*) Let a *task* be a sequence of instructions that we would like to run on one thread. A task in our manuscript is atomic: Once started, there are no further dependencies on other tasks unless these tasks are children tasks, that have to complete before we continue.

The control flow graph (Sect. 7.2) of a problem helps us to identify tasks. Yet, the decomposition of a problem into tasks is seldom unique. A good task

- is a logical unit (does one thing only); and/or
- can run in parallel to other tasks; and/or
- is reasonably small; yet
- combines enough instructions such that the overhead to launch a task relative to its compute cost remains reasonable.

These criteria are wishy-washy. In the end, the decision which instructions to combine into one task (*task granularity*) is up to the programmer.

Once we group our control flow graph into sequences or blocks of instructions and thus identify tasks, these tasks inherit the data dependencies: If a task A hosts an instruction a and a task B hosts b and there is a data flow dependency from a to b. Task B consequently has to be executed after task A.

Definition 17.2 (*Task graph*) A task graph is a *directed acyclic graph* (DAG): Each task (node in the DAG) needs a clear set of preconditions represented by incoming edges, while the graph over the tasks may not have any loops.

Loops in the control flow graph thus require us to break down the way we construct our task graph into multiple steps: Whenever we create a task graph (*task graph assembly*), it has to be a DAG, and we cannot map a loop directly onto a DAG. We might therefore assemble one task graph per loop iteration, or we might unroll the loop before we create the task graph. A task graph is a snapshot of the loop behaviour.

In this context, we emphasise that a task graph does not have to be *static*: We can always dynamically add tasks to a task graph. The only constraint is that the graph has to remain acyclic. We can only add further children to existing tasks or further follow-up tasks that either have dependencies on previous steps or not. This constraint helps us to formalise how tasks are deployed to the threads of our system:

Definition 17.3 (*Ready task*) A *ready task* is a task which can be computed as all input data are available.

Once we have grouped our code logically into tasks, the tasks (including their dependencies) are thrown into a *task pool*. Individual threads grab ready tasks from this task pool and execute them. Throughout this execution, they might add more tasks to the task pool. As soon as a thread has terminated, it grabs the next ready task. As the termination of a task implies that data feeding into follow-up steps is computed, i.e. data flow dependencies are fullfilled, each termination may unlock further tasks, i.e. make them ready.

As we work with a DAG, there is always at least one ready task as long as there are tasks in the pool.

The execution scheme describes a work stealing mechanism. However this dynamic scheduling approach is now phrased in terms of activities (tasks) as modelled by the programmer rather than data regions (cmp. Sect. 16.4). The number of ready tasks within the task pool is a lower bound on the concurrency level of the code: If tasks themselves are parallelised—as they host a parallel for, e.g.—then the concurrency level of the overall pool is even higher than the ready task count. It is up to the OpenMP implementation to decide whether they work with a centralised task pool or a task pool per thread where task stealing occurs if and only if a local pool runs empty.

17.2 Basics of OpenMP Tasking

We introduce our OpenMP task ingredients with the following code snippet:

```
  #pragma omp parallel
{
  #pragma omp single
  {
    #pragma omp task
    {
      std::cout << "Task A" << std::endl;
      #pragma omp task
      {
        std::cout << "Task A.1" << std::endl;
      }
      #pragma omp task
      {
        std::cout << "Task A.2" << std::endl;
      }
      #pragma omp taskwait
      std::cout << "resume A" << std::endl;
    }
```

```
    #pragma omp task
    {
      std::cout << "Task B" << std::endl;
    #pragma omp taskwait
    #pragma omp task
      std::cout << "Task C" << std::endl;
  }
}
```

Lets run through the snippet step by step:

1. The `parallel` section fires up our threads. Lets assume we have four threads (`OMP_NUM_THREADS` is set to four). Prior to the pragma, only one thread, the *master thread*, is alive. We hit the pragma and fork the other three guys. This is the fork-join model. We now have four threads up and running.
2. OpenMP follows a SPMD model, i.e. all threads run the same code. We next produce a couple of tasks. However, we don't want all four threads to produce tasks, so we tell OpenMP that we want only one thread to continue. This is achieved through `single`. We could have used `master`, alternatively. It has the same effect yet ensures that the following code section is only executed on the master (first/original) thread. In our example, we do not care: Any thread can run the following code, as long as only one does so.
3. We create a task (lets call it A as it plots `Task A` to the terminal) and add this one to the task pool. Skip the remainder within task A's brackets and notice that we immediately create a task B, too.
4. Once A and B are put in the task pool, or original task (let's call this task S due to the `single`) waits for A and B. They have to terminate before S can continue. This is achieved through the `taskwait`.
5. After both have terminated, the task S yields another task C. C is a one-liner as there are no block brackets. No further tasks are scheduled anymore. However, the closing brackets of the `parallel` section implicitly make us wait for all tasks to terminate, so S implicitly waits for task C.
6. If a thread grabs task B, it prints `Task B` to the terminal. After that, the task is complete and the thread tries to grab the next one from the pool.
7. If a thread grabs task A, it prints the message `Task A`. It then spawns another two tasks A.1 and A.2 and immediately after that waits for them. Once they are complete, it dumps the resume message and thus completes the task's instructions.
8. We finally join all threads again.

One possible execution pattern is illustrated in Fig. 17.2.

OpenMP Single and Master

OpenMP `single` and `master` regions are executed only on one thread. They may not be nested within a worksharing OpenMP region.

17.2.1 The Task Dependency Graph and Ready Tasks

Creating (spawning) tasks and inserting waits both tell OpenMP which task dependencies do exist. To the right in Fig. 17.1, I illustrate the task dependencies as the OpenMP runtime system might see them at a certain point: It recognises that our program has produced four tasks so far. Three of them are ready, as they have no incoming dependencies, two of them cannot continue/start at the moment.

1. The code's structuring into `task` blocks defines the task control graph.
2. Whenever we make a snapshot of the system's state, we can write down all the existing tasks (both those assigned to threads and those pending in the task pool) plus their dependencies. This is the task DAG. In our example it is always a tree, but we will generalise this one later. The tasks without any incoming dependencies in this snapshot are the ready tasks, i.e. those tasks that an idling thread can grab and process.
3. Every `task` statement creates a new node within this task graph.
4. Every `taskwait` adds fan-in dependencies to the task graph.

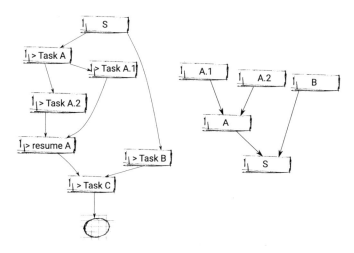

Fig. 17.1 Task control graph of the code snippet (left), and task dependency graph at a certain point (right)

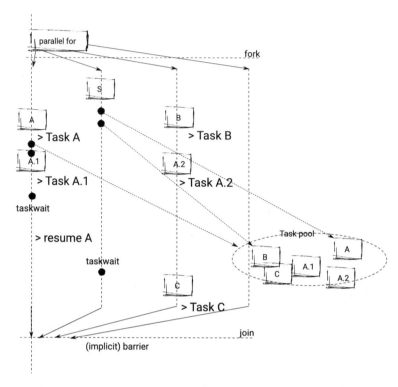

Fig. 17.2 Execution pattern of example source code from Sect. 17.2: The single environment S it grabbed by the second thread. Thread S creates two tasks which are immediately grabbed by thread one and three such that thread two hosting S has to wait for their completion. When task A spawns A.2, the runtime decides not to put it into the task pool but instead immediately processes this one on the current thread as it hits the taskwait instruction. Dotted lines illustrate task creation, i.e. throwing tasks into the task pool

5. If a task terminates, its dependency edge in the graph of tasks managed by the runtime system is removed. The task graph thus hosts both a subset of the tasks that exist overall and a subset of the dependencies. It changes over time.
6. If a task spawns another task, both tasks immediately are ready (if we read the code step by step). If the creating task (the one that spawns) issues a synchronisation (either via `taskwait` or as it closes the `parallel` section), it adds a task dependency i.e. becomes waiting and thus degenerates to non-ready.

OpenMP's task graph construction is always *dynamic*, i.e. your code tells OpenMP step by step what the task DAG looks like. Since you build up the task graph incrementally, OpenMP can start to distribute and execute tasks while you still build up the overall graph.

Scheduling Point

The decision which ready task to process next is up to OpenMP. OpenMP however relies on *scheduling points* in your code, i.e. steps where the OpenMP runtime source code is allowed to "think" about the next steps. It does not arbitrarily or parallel to your run make further scheduling decisions. OpenMP implicitly reaches a scheduling point whenever a task terminates. It is then up to the runtime to pick the next one. Furthermore, OpenMP regards any task creation point as scheduling point. That is, directly after the programmer creates a task, it is up to OpenMP to decide with which task to continue. Finally, any type of implicit or explicit synchronisation provides a further scheduling point: If you close a parallel region or if you issue a `taskwait`, the scheduler takes over.

17.2.2   Task Synchronisation

Definition 17.4 (*Task synchronisation and orchestration*) Task *synchronisation* is either done explicitly through OpenMP pragmas or implicitly at the end of scopes. Once a task is running, it is not interrupted unless you make it synchronise explicitly.

In OpenMP, we hence have three types of task synchronisation mechanisms:

1. *Explicit barriers* are introduced by the keywords `taskwait` and `taskgroup`. If a code runs into a `taskwait`, the current thread is suspended until all direct children have terminated. The `taskwait` does not wait for the tasks spawned by direct children tasks, i.e. it does not wait for descendents!
   The statement `#pragma omp taskgroup` is typically followed by a block (use brackets). At the end of the taskgroup, i.e. when these brackets close, the current thread waits for all direct children plus all of their descendents. That is, if the children recursively have spawned further tasks, then we wait for these, too.
2. *Yielding* means that a task suspends its execution and gives the runtime the opportunity to continue with another task. To ensure that the multitasking overhead remains low, OpenMP typically processes tasks en bloc, i.e. once a ready task is grabbed by a thread, it is also completed. With `#pragma omp taskyield`, we can alter this behaviour.
3. *Implicit barriers* are barriers as above, but we don't write them down explicitly. There are three major types of implicit barriers: At the end of the parallel region, we have an implicit barrier. To be able to join the threads, all tasks have to be completed, too. At the end of a task group, we have an implicit barrier on all descendents. There is also a task loop construct in OpenMP. It behaves similar to OpenMP's `for` but explicitly creates tasks; which have a slightly higher overhead, different default visibility semantics (see below) but are grabbed by threads rather than scheduled. All implicit barriers besides the ones for the overall parallel section can be removed by adding a `nowait` clause to the OpenMP pragma.

Task Interruption/Deferral

Tasks are usually not interrupted by OpenMP unless they yield explicitly. However, the OpenMP runtime is allowed to *defer* a task, i.e. to interrupt it to continue with a direct child task. The original task resumes typically on the same hardware thread unless it is explicitly made migratable. This ensures that not too much data has to be moved between cores.

From a programmer's point of view, the OpenMP scheduler kicks in whenever we spawn a task and every time a task terminates; unless we yield. However, this knowledge in practice is of limited use, as we have limited control over the scheduler's policy. In particular, there is no fairness guarantee.

Definition 17.5 (*Starvation*) *Starvation* means that a task is ready yet is never executed, as the program always spawns further tasks and the scheduler makes them "jump the queue".

From the outside, starvation is difficult to distinguish from deadlocks.

To ensure that the task pool isn't overflooded with tasks, OpenMP can at any point decide to ignore further spawns and to execute new tasks straightaway. Logically, it defers the parent task (puts it on hold) and makes the thread running the deferred task process the children directly. In some cases, children might be very expensive: It might be a good idea to suspend a parent task at one point and to continue to work on its children directly, but then later some idle threads should take over, re-grab the deferred parent task and continue. Usually, OpenMP does not allow this, i.e. once a thread has taken a task, it runs it. If it is temporarily on hold, it remains tied to the thread. If another thread should have the opportunity to step in, you have to mark your task as untied.

17.2.3 Variations of the Example

- If we removed the single from our toy code, the whole task graph is built up OMP_NUM_THREAD times.
- If we removed the first taskwait instruction, then task A spawns A.1 and A.2 but does not wait for them. It might terminate before A.1 and A.2 terminate. There is a second taskwait if we scroll down. This one makes the single thread spawning A and B wait for A and B before it spawns C. As soon as we comment out the first taskwait, task C might run parallel to A.1 and A.2.
- If we removed both taskwait annotations, all tasks could run in any order.
- If we removed the first taskwait annotations and replaced the second one by a taskgroup, then the taskgroup would wait for A and B plus A.1 and A.2.

17.3 Task Properties and Data Visibility

By default, all tasks have the same priority, i.e. we do not distinguish which one is important and which one is not.

Depth-First Task Graph Traversal

Most task runtimes traverse the task graph in a depth-first order, i.e. if a scheduler has multiple ready tasks to choose from, it typically selects a direct child of a previously running task. The rationale is that a task spawned by another task is likley to operate on similar data, i.e. some data might already be in the registers or a close-by cache.

If you want to impose a priority on your task, you can annotate the spawn mechanisms with the `priority(n)` clause. `n` is an integer. The higher the value the more important the task.

In principle, tasks fit to the BSP mindset we have used to introduce classic OpenMP. All data are shared, threads are forked and joined, and the same synchronisation mechanisms as for BSP are available. There is a delicate difference however when it comes to the default visibility of variables. OpenMP has three categories of variable visibilities:

- *Predetermined data sharing* is a data sharing rule that derives directly from the usage. Our loop iteration counters and ranges within a `parallel for` for example are predetermined: You know that they are thread local and you cannot change this behaviour (you can still explicitly write down their visibility but it has to match the default, i.e. there's not really any freedom).
- *Explicitly determined* visibility means that you write down `private`, `first private` or `shared` in the pragma.
- *Implicitly determined* means you skip any explicit annotation and you rely on the default rules.

In a classic loop context, `shared` is the default visibility for implicitly determined sharing. For tasks, this does not make sense: If a task is spawned, it sees all the variables defined "around it" at the moment it is created. In the absense of a task wait mechanism, we know that these variables might already be dead when the task launches: Let the parent create a local variable, spawn a new thread and then terminate. The termination of the parent might mean that the local variable is long gone by the time the spawned task is grabbed by a thread and actually executed. Therefore, OpenMP has altered the default visibility:

Definition 17.6 (*Default visibility*) For a task, variables are by default `firstprivate`.

If you label variables as shared or you work with pointers, then multiple tasks still can access the shared memory as kind of a shared scratchboard. As a result, you might still produce data races. Yet, all the shared memory synchronisation mechanisms continue to work. That is, you can use atomics, and you can use critical sections to orchestrate memory access.

17.4 Task Dependencies

Our task mechanisms so far imply that the task pool (snapshot of the task graph) always hosts trees. Tasks either are ready, or they wait for their children to terminate. OpenMP gives you the opportunity to model more complex task graphs. For this, it simply uses memory locations, i.e. memory addresses:

- If you annotate a task with a `depends(out:x)` clause, OpenMP internally keeps note that your task will write something into the memory location of x when it terminates.
- If you annotate a task with a `depends(in:x)` clause, OpenMP will analyse which existing tasks write to x, i.e. have an out dependency. These tasks will have to terminate before this task becomes ready.
- The annotation `depends(inout:x)` is a shortcut for a task which reads from a memory location and then writes to this one again.

In principle, the construction of a task dependency graph in OpenMP is thus very simple—in particular as you can combine an arbitrary number of dependencies. However, you have to keep in mind that it is OpenMP's philosophy that program annotations are augmentations, i.e. any OpenMP code should continue to run even if OpenMP is not present. Therefore, you have to ensure that any out-dependency is defined (the corresponding tasks are spawned) before you use the respective address as in-dependency. If an in-dependency is not matched by a previous out-dependency, OpenMP considers this an *anti-dependence*.

Definition 17.7 (*Anti-dependence*) An *anti-dependence* means that task A preceedes task B, that task A reads from a memory location x, and that task B writes to a memory location x.

An anti-dependence means that two tasks A and B have to be serialised. Otherwise, OpenMP might break the code semantics.

Some important material

- OpenMP Architecture Review Board: *OpenMP Application Programming Interface—Examples*, Version 5.0.0, 2019 https://www.openmp.org/wp-content/uploads/openmp-examples-5.0.0.pdf

> **Key Points and Lessons Learned**

- OpenMP relies on a task pool into which tasks can inject further tasks.
- Task dependencies are built up through a wait clause or by explicit depends instructions.
- Tasks synchronise explicitly, implicitly, or they yield.
- Dependencies model dependencies, not data flows. If we need information flow between tasks, we use the shared memory space. Yet, the default visibility is firstprivate and we have discussed why. For information flow between tasks, we can use critical sections or atomics.
- If our task graph resembles BSP, we should, despite all the nice task features, use the more traditional, BSP-style pragmas. They are faster.

GPGPUs with OpenMP

18

Abstract

GPGPUs can be read as multicore chips within the compute node which host
their own, separate memory and multiple cores which are slow yet nevertheless
extremely powerful. This power results from SIMD on a massive scale. We give
a brief, very high level overview of the GPU architecture paradigm, before we
introduce OpenMP's target blocks. The data movement between host and accelerator requires dedicated attention, while a brief discussion of collectives, function
offloading and structs (classes) for GPUs close the discussion.

Assume that a house is on fire, the next river is a mile away, and there is only
one fire fighter who can get close enough to the fire so he can throw water in.
This water is, in our thought experiment, delivered in buckets.

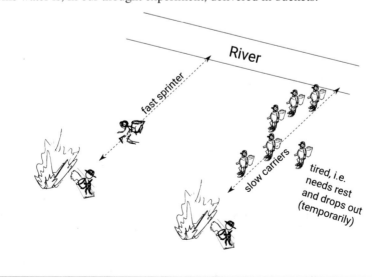

You have to make a decision how you provide this one fire fighter with buckets: Do you hire one sprinter, i.e. a guy who can run really fast, or do you rather hire an old-men's football team, i.e. a bunch of slow guys?

- What is the better (faster) solution?
- Which one is better from a Dennard scaling point of view?
- What happens if one carrier suddenly becomes tired and needs a short rest or a convenience break?
- What happens if the sprinter suddenly becomes tired and needs a short rest?

Standard CPUs are the racehorses of computing. They are fast and versatile. GPUs have been designed exactly the other way round: They are relatively slow and comparably simplistic (cmp. our discussion on lockstepping in Sect. 7.3.2). However, they can process vast amounts of data and instructions in a very homogeneous way in parallel.

Historically, they are designed for computer graphics/games where the same operation has to be done for every pixel. If all pixels are processed in parallel, the individual operations are allowed to be relatively slow, as long as the overall throughput is high. We know that such a uniform processing scheme maps efficiently onto vector processing.

The scientific computing community has a large interest in GPU architectures, too. Uniform (vector) operations are omnipresent, and the community suffers from Dennard scaling, i.e. needs to get the power consumption (frequency) down. General purpose graphics processing units (GPGPUs) emancipate from graphics and offer vector processing capabilities in a novel way. They are additional devices that plug into your computer and can do graphics but also general computations. Some modern GPGPUs don't even have a graphics port anymore, i.e. they are compute-only devices.

Definition 18.1 (*Host and devices*) We distinguish *host* and *devices*. The host is the main compute unit (the CPU) which orchestrates other machine parts, i.e. all the GPGPU accelerators.

With the increasing importance of GPUs for the mainstream (super-)computing market and the increasing capabilities of GPUs, some developers started to articulate their frustration that they effectively have to write two codes when they deal with GPUs: The main code for the CPU (host) and then the functions (kernels) that they want to run on the GPU (device). While the programming language CUDA continues to dominate the GPU software market, more and more developers favour a pragma-based approach. It allows them to write all of their code in one language, while they annotate those parts that are a fit for a GPU. A compiler then can take care of the data transfer and how to generate the right microcode for the right GPU. Furthermore, the code continues to run "as it is" on a CPU without an accelerator.

There are two important pragma dialects for GPUs: OpenACC and OpenMP. OpenACC is tailored towards accelerators (hence the acronym). Though it resembles

OpenMP, it is completely independent and realises a few different ideas (programmers for example can work declarative rather than prescriptive). As we already have used OpenMP, we stick to OpenMP for the GPGPUs.

18.1 Informal Sketch of GPU Hardware

The omnipresent term in GPU programming is CUDA. Released in 2007 by the company NVIDIA, CUDA stands for two things:

Definition 18.2 (*CUDA*) CUDA stands for *Compute Unified Device Architecture* and is a name, on the one hand, for a particular programming language. On the other hand, it circumscribes the architecture of a GPU chip, i.e. the principles of how to build an accelerator.

In this manuscript, we ignore the CUDA language, which is a language that is very similar to C yet sufficiently different to make it a topic of its own. Yet, we should have a rough understanding of how a GPU works. Though NVIDIA is not the sole vendor of GPUs, CUDA remains the blueprint of a GPU hardware design.

18.1.1 Memory

A CUDA device is an accelerator, i.e. an additional computer within the computer. A program can be launched by the CPU on the GPU, but it is a different machine inside the machine. Typically, CUDA programs are not full-blown applications, but particular functions. The program or each function, respectively, is called a *kernel*.

The accelerator has its own memory, i.e. CPU and GPU memory are physically separated. If the GPU requires data from the CPU, we have to copy it.

Definition 18.3 (*Unified memory*) Modern CUDA architectures offer *unified memory*. In this case, a compute kernel can (logically) access the CPU's memory (and vice versa). We work with one large memory address space. Technically, the two memories are still separate and we have *on-chip* memory access and *off-chip* access. If the GPU faces a CPU memory request, it issues a page miss, i.e. it internally copies data over. To allow the hardware to release your software from the cost to transfer data, you have tell your system that you want it to store data in this unified memory when you allocate the data. NVIDIA offers specialised versions of `new` or `alloc`, respectively, for this.

Originally, GPGPUs—I will use GPGPU, GPU and graphics cards as synonyms from hereon—have been totally separate from the CPU. With improving integration, i.e. more powerful kernel capabilities and better memory integration, this distinction blurs.

We do not dig deeper into this idea of a logically shared memory and the integration, but instead copy data forth and back. This can either happen explicitly, i.e. we transfer the data, or the GPU runtime and compilers can derive the required data movements from source code and trigger data movements. We call the latter approach *managed memory*. Managed memory versus manual data transfers are a decision we have to make as programmers.

18.1.2   Streaming Multiprocessors

Each GPU card hosts multiple processors. It can be seen as a MIMD multicore machine, as all the processors have access to the common GPU memory. Each processor has its own control unit, i.e. runs its instruction stream completely independent from all the other processors. If two processors communicate with each other, one processor has to write data to the shared memory from where the other one can grab it.

On state-of-the-art GPUs, we have around 80 of these processors. They run at relatively low speed (typically way below 2GHz). The massive power of GPUs thus has to come from somewhere else. Indeed, the GPU processors are called streaming multiprocessors (SM) in NVIDIA's CUDA terminology as they are particularly good in streaming data through the chip. AMD calls them computer unit (CU). Both rely on a particular SIMD design flavour within their cores:

18.1.3   Core Design

Each GPU processor hosts a certain number of threads. NVIDIA today calls them CUDA cores. AMD speaks of "parallel ALUs" which is closer to the terminology we have used so far. 32 or 64 are popular numbers for these CUDA cores or parallel ALUs. With 80 cores and 64 threads, a GPU offers up to 5,000 CUDA cores in total.

These CUDA cores are a mixture between hardware threads and full-blown cores. They have their own registers and ALUs, but they cannot work independently. All the cores within a SM operate in lockstep mode; even though recent GPU generations weaken this constraint. The cores are independent, but they share the control logic. The CUDA cores per SM are CUDA's materialisation of vector computing.

There is one further difference compared to standard CPUs: The register count within each CUDA core—and therefore SM—is massive. More than 200 is typical. If an instruction on a CPU thread has to wait for data from the memory, the CPU has two options: It can either wait or it can save all register content to fast nearby memory that we call cache, load the registers of another (logical) thread and continue with this one. We call this thread switches. They are expensive.

GPUs organise their code in *warps* (NVIDIA's name) or *wavefronts* (AMD). I use warp most of the time for historical reasons. A warp is a set of 32 or 64, respectively, logical threads in the source code. As discussed, the instructions of these logical threads run in lockstep mode. AMD calls them work item. In a CPU world, the bundling (lockstepping) of vector instructions introduces a problem: If one of them

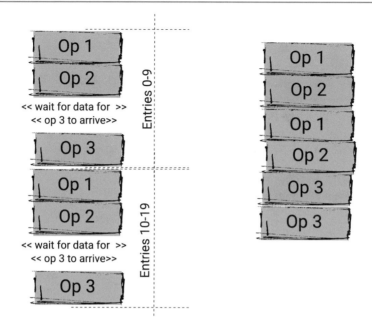

Fig. 18.1 Left: Assume an array of 20 entries is split into two chunks. Each chunk is handled in parallel via SIMD instructions. We run three SIMD steps on the array entries. Unfortunately, the third operation requires the CPU to wait for some data to drop in. If we had two cores, we could run the two chunks in parallel and each chunk would have to wait. We don't have two cores here, so we deploy both chunks to the same core. We oversubscribe it. The chunks are scheduled one after the other as the chip recognises that saving all register content after operation two would last longer than the actual wait. Right: A GPU architecture relies on oversubscription. Different to the CPU, it can switch between the warps, i.e. chunks, very quickly. There are enough "spare" registers to hold all temporary results from operation two. Should it have to wait for some data for the third operation, it can consequently switch to another chunk and continue to work with this one while the first chunk "waits" for data to drop in

has to wait for input data, the remaining have to wait, too, or we have to thread switch.

On the GPU, we switch threads all the time: If one of the 32 lockstepping threads has to wait as data is not there yet, the GPU can simply switch to another warp, i.e. let another set of threads progress. The high number of registers makes this possible. If the GPU code is not extraordinary complex, there is no need to back up register content, as we have so many registers per hardware thread. Up to 64 of the warps can run "in parallel": they do not really run parallel, they simply are active, i.e. it is up to the GPU to decide which of the 64 active warps to progress (Fig. 18.1) at any point. Thread switches are a design principle rather than a performance flaw. NVIDIA calls the set of warps that are currently active on a GPU processor a *thread block*, while AMD speaks of a *work group*.

Definition 18.4 (*Warp interleaving*) Let multiple warps (sets of threads) be active on an SM. Some of them might be ready to run the next instructions, others might wait for data to arrive. The SM cannot run the independent warps in parallel, but it can always pick a warp that can progress without any further wait. It effectively *interleaves* the execution of different thread sets (*warps*).

MISD

We can interpret warp interleaving as a realisation of MISD in the tradition of pipelining.

Today, NVIDIA and other vendors make their cores flexible. New GPUs for example have a different program counter per CUDA core and not one for all threads of a warp. Therefore, they become more resilient against thread divergence. However, high thread divergence leads to high scheduling overhead and any synchronisation of "unlocked" threads remains expensive. The principle therefore remains: GPUs perform best if

- 32 or 64 threads run in lockstep mode, i.e. in a SIMD way—though we call it Single Program Multiple Threads (SIMT) here to highlight that we are working with something more similar to a thread than a SIMD register;
- several sets of threads (warps) should be able to progress at any time per core. As a rule of thumb, at least four active warps give good performance.

GPUs are hardware that is built for codes with (i) a significantly higher concurrency than what we find in the hardware and (ii) infrequent synchronisation needs.

Definition 18.5 (*Occupancy*) The *occupancy* of a GPU processor is the number of active, running warps on a processor divided by the maximum number of active warps a processor could handle.

If you have a code with an occupancy of one, the GPU has no more resources to take up another active warp. If it stalls (as data is for example not available yet), then there is nothing the chip can do to mitigate this problem. The term highlights the capabilities of the GPU to switch warps efficiently.

Occupancy versus concurrency versus parallelism

Concurrency is a theoretical quantity, and parallelism is something you observe. Occupancy correlates the two quantities. Occupancy in combination with the thread divergence (cmp. the masking and blending discussion in Sect. 7.3.3) determines how fast your code performs on a lockstepped machine.

The occupancy plus the thread divergence give you a theoretical estimate of how efficient your code can run on a GPU. If you struggle to meet full occupancy, your code might be busy with data transfer, might be too complex and thus exceed the number of registers on the GPU, i.e. further warp interleaving is not possible, or it might have insufficient parallelism:

To really exploit a GPU, we should have a concurrency of at least

$$\underbrace{80}_{\text{SMs}} \cdot \underbrace{4}_{\substack{\text{active warps; can be up to 64}}} \cdot \underbrace{64}_{\substack{\text{warp size (threads)}}} \tag{18.1}$$

and it should be concurrency in a BSP/SIMD sense at least on the warp-level.

Multicore CPUs are designed to have as many independent cores as possible. Their architecture focuses on the control flow. GPUs are designed to stream as many instructions as possible with no delay between two streams through the core. Their design is data-centric.

GPUs are good for dense linear algebra like vector-vector or matrix-vector multiplications. New GPUs go one step further: They provide tensor cores on top of the SMs. These are further specialised cores which do matrix-matrix multiplications where every single step is optimised in a SIMD sense; but in hardware!

20,480 active threads in (18.1) is quite a lot. With this number in mind, it is clear that not every algorithm is a fit for a GPU. If, however, an algorithm exhibits such a concurrency level and large parts of this concurrency is SIMD-style, then GPUs can give a massive boost as their enormous concurrency pairs up with a no-wait execution due to warp interleaving.

Our discussion will now move on to discuss how to write proper GPU code if the algorithm is a fit. This means we have to express the SIMT versus SM properties, and we have to transfer the data to the GPU as well as to retrieve it again.

Architecture review

In a way, GPUs add the von Neumann bottleneck to our code once more: Data has to be brought from the CPU memory to the right CPU and from the GPU memory to the SM. On top of that, data has to be moved into the accelerator and has to be fetched back.

18.2   GPU Offloading

Creating a compute kernel, i.e. a program that could run on the GPU is close to trivial with OpenMP:

Definition 18.6 (*Target*) A *target region* in OpenMP describes a block which is executed on a device.

Once we wrap a code block with

```
#pragma omp target
{
  ...
}
```

the compiler is told to create machine code for the GPU. Furthermore, the target instructs OpenMP to kick off the block's instructions on the GPU, to wait for its termination, and then to continue. We call such a behaviour *blocking*, as the host core is blocked while the GPU is busy. You cannot nest targets.

Data transfer in speedup models

For a fair modelling of GPU performance, it makes sense to read the data transfer prior and after a GPU kernel as sequential code part f in an Amdahl/Gustafson sense. This data transfer overhead will depend on the problem size.

We formalise this behaviour: The target first opens a *stream* to the GPU. The term stream here means that data is streamed to the accelerator. It moves data. Next, the target pragma creates one new team:

Definition 18.7 (*Teams*) A *team* is a league of threads.

If we write down omp parallel, we establish a team on the host and fork the threads. If we write down omp target, we create one team on the device and fork threads. The team will be mapped onto one SM:

Target

A target (without any further teams modifications) creates a new virtual processor within OpenMP that behaves in the same way as the original multithreaded processor. It is up to OpenMP do roll this new "processor" out to an accelerator.

In contrast to parallel, code within a target construct continues on the master thread of the newly created team only. target on the GPU behaves similar to parallel immediately followed by master on the host. The default of a team is that the team's code runs serially!

Within the new team, we can now use `parallel for` to run through a loop in BSP style; though technically all threads of a team exist right from the start and are just masked out prior to the `parallel for`. Each team hosts its own thread pool. That is, within a team, we can use dynamic loop scheduling, and we could even try to port a whole multitasking environment onto the GPU—though it really depends on the hardware and software ecosystem in your machine to which degree this is supported (efficiently). In general, classic BSP with low thread divergence will perform best.

On most systems, you cannot specify how many threads you want to use within a team. This depends on the hardware. Therefore, there is also no `OMP_NUM_THREADS` environment variable for our "virtual" processor on the GPU. You can however specify the maximum thread count via `thread_limit`. If this is not a multiple of the SM's CUDA core count, then you will leave threads idle.

It is up to OpenMP how to distribute the threads within a team over warps.

Nested blocks (1/2)

You cannot nest a `target` within another target block, but you can use targets within a parallel block and parallel blocks within a target.

Nested blocks (2/2)

Theoretically, you can use a `target` block with a nested `parallel for` which is not strictly nested. In principle, you can also nest multiple `parallel for` loops in GPU kernels. However, we have observed that many OpenMP compilers struggle to parallelise properly in such cases. In the ideal case, there is one parallel loop strictly nested within your target, and the `target` and the `parallel for` pragma are written down as one combined construct.

18.3   Multi-SM Codes

While `target` marks code that should go to the device, you have to explicitly tell OpenMP if you want to use more than one team. This is achieved via OpenMP `teams`. The statement can be augmented with a `num_teams` clause. With the clause `thread_limit` you can specify a maximum number of threads per team. OpenMP however might choose a smaller number. In any case, the number of threads per team will be the same for all teams of a league:

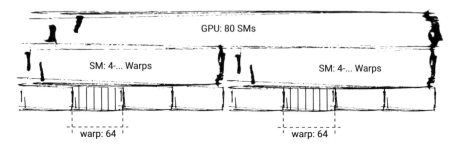

Fig. 18.2 With `teams distribute parallel for simd`, a large for loop is first split up (distributed) into teams. This splitting is static as each team is mapped onto one SM and the SMs do not synchronise or load balance between each other. Once a team is given a certain loop range, it has to process this loop range. The `simd` statement splits the per-SM loop range into vector fragments which keep all SM threads busy. They form warps. The final set of warps is progressed by the SM. As it is labelled with a `parallel for`, some load balancing is realised by the SM. Yet, this load balancing is realised through warp interleaving

Definition 18.8 (*League of teams*) A `teams` block creates a *league of teams*, i.e. a set of teams where each team has a distinguished master thread. OpenMP's `parallel` statement creates the host thread team, while `target` without a `teams` qualifier creates one device team.

Teams do not really add something new that we haven't been able to express from a programmer's point of view. The essential rules around teams are straightforward:

- We may not branch in and out of a team.
- The `teams` construct may only occur within a parallel region or a target.
- If `teams` occurs within a target region, then no other statement may occur outside of the team, i.e. the `teams` statement has to follow the target immediately. You cannot have something within the target block that is not not covered by the teams too. Teams have to be strictly nested (Definition 16.4 from Chap. 16).[1]
- Within a team, we may only have a parallel construct, or a SIMD construct, or a distribute construct.

With the CUDA architecture in mind, it is obvious why we may not have any instruction before or after a `teams` block within the same target: Teams are made to allow a compiler to map logical threads onto different SMs. As SMs are more or less totally independent of each other, we want to launch whole kernels on them. We don't want one kernel on one SM affect the other SMs. As teams are so restrictive, I typically always use them in combination directly with a target, i.e. instead of opening a target block and then immediately a teams after that, I write `omp target teams`. Even more than that, I combine it with `distribute` directly:

---

[1] This constraint is weakened starting with OpenMP 5, but it makes sense, for the time being to assume that it holds in the strong form postulated here.

The distribute construct is the parallel for of teams: It takes the subsequent for loop (which has to follow all rules we had set up for the parallel for loops before) and splits up the iteration range between the teams. Though teams accept a clause dist_schedule, the only distribution schedule allowed here is static. That is, teams cannot load balance between them. This is exactly what allows OpenMP to map them efficiently onto warps and tasks (Fig. 18.2).

We conclude that a popular OpenMP GPU pragma looks similar to

```
#pragma omp target teams distribute parallel for simd
```

which reflects the hierarchical decomposition of a loop over the GPU, its SMs and (CUDA) cores on the SMs. It also illustrates that the distribute statement only distributes the loop range over the team's master (or initial) thread. It is the parallel for which then launches the per-team parallel region and the further loop decomposition.

Multi-SM computations

Each target region is a task in an OpenMP sense and specifies one compute kernel. Modern GPUs can run multiple compute kernels simultaneously. If you cannot occupy all of your SMs via one compute kernel, nothing stops you from launching different targets on your GPU at the same time.

18.4   Data Movement and Management on GPUs

The task analogy for GPU targets explains why scalars within OpenMP targets are by default firstprivate. Whenever OpenMP launches the compute kernel on the GPU, it copies scalars automatically over. The GPU task is ran—so far immediately— and has all data locally available.

firstprivate GPU semantics

Scalars within GPU kernels are firstprivate.

Copying scalars is cheap. Big arrays are a different story. They are not copied automatically.

18.4.1   Moving Arrays to and From the Device

For big arrays, the programmer has to instruct the compiler which (parts of the) arrays go to the GPU, remain there, and which have to come back with updated content. By default, arrays are mapped onto a pointer (array) on the device of size zero. That is, your code might compile if you forget to specify how an array is transferred. The first access to such an array however will cause a crash, as the array's pointer points to zero and to a field of length zero.

Programmers have to annotate the target environment with data movement clauses.

1. map(to:x) instructs the compiler to create an array x on the device. The array is initialised with firstprivate semantics. It is copied from the host to the device before the kernel starts to run.
2. map(from:x) instructs the compiler to create an array x on the device, and to copy its content back once the kernel has terminated.
3. map(tofrom:x) combines both clauses.
4. map(alloc:x) tells the compiler to create an array x on the device. This array is neither initialised nor is its content copied back once the kernel has terminated.

In line with OpenMP's vision that codes should still be correct if we translate without OpenMP/GPU features, all the clauses assume implicitly that the array x exists on the host if there is no GPU support. The four clauses are all you need if you handle arrays where their size is known at compile time. However, if you have a function which accepts pointers to dynamic memory regions, OpenMP cannot know the size of the underlying data structure.

Definition 18.9 (*Arrays on devices*) If you handle *arrays on your device*, you have to map them explicitly. If the array is dynamic, you also have to instruct OpenMP which regions of an array have to be mapped.

If you use dynamic arrays, the data transfer clause has to be augmented by further range specifiers.

```
const int L = 20;
double* myArray = new double[L];
#pragma omp target map(tofrom: myArray[0:L])
{
   ...
```

This means that you can also synchronise only subranges of an array, as you specify the offset and the number of items to be copied forth and back.

18.4.2 Data Management and Explicit Data Movements

We dive into OpenMP's data movement realisation to clarify that it works internally with some reference counters: For all of our GPU code, OpenMP hosts one table per GPU. For our purpose, we may assume that the table has three columns: a column with (CPU) pointers, a column with GPGPU pointers, and the right column is an integer counter. The table maps each pointer from the host onto a memory on the GPU and it tracks its lifetime (Fig. 18.3).

When you translate, the compiler runs through your GPU code, i.e. your target block, and analyses its data reads and writes: For each access to a certain variable within a block, the target's kick-off code that our OpenMP compiler adds automatically looks up whether the GPU knows the address of this variable, i.e. whether there is a row in the table which maps the CPU address onto a GPU address. If not, the generated code allocates a new memory location on the accelerator, adds an entry of its address to the GPU table, and copies the data over from the host into the newly generated GPGPU memory. Once copied, the data reference counter is set to 1. All memory accesses to the respective variable on the accelerator are now replaced with the GPU memory location from the table. This does not work for data of dynamic range. In this case, the compiler requires you to copy stuff manually via map. The kernel itself creates a mapping entry for any memory access but makes it point to 0 on the GPGPU. Your explicit map statement allocates the memory on the GPU and sets the counter to 1. Once we know that the counter is at least 1, we can copy the data from the CPU to the GPU (if wanted by the user). In the same way, the compiler can generate an epilogue which copies GPU data back to the CPU if instructed to

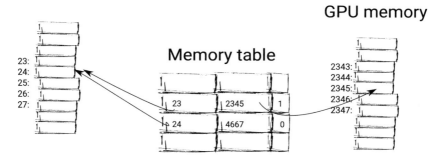

Fig. 18.3 The memory of CPU and GPU is (logically) totally separated. I denote this through different addresses here. OpenMP's memory table holds an entry (23, 2345, 1) which means that the CPU memory location 23 is mapped onto the GPU's memory location 2345. The 1 indicates that this mapping entry is valid. The mapping from 24 onto 4667 is not valid anymore as the GPU memory is already freed and might be used by some other GPU kernel

do so. When we leave a target block, the counters of all the variables that are used at least once within the target block are decremented.

We end up with an address replacement process: OpenMP generates two versions of a target block. One for the GPU, one for the CPU. The CPU version works with the real memory addresses. In the GPU version, all memory addresses are replaced with their GPU counterpart.

Smart pointers and garbage collection

The data bookkeeping as sketched realises the principle of a smart pointer: We count per variable how many references to this data remain alive. If the counter reduces to zero, no one needs the data anymore and we can free the memory space. This is one way to realise garbage collection.

Making the reference entry an integer allows us to launch multiple GPU kernels concurrently. If a pointer for a CPU memory region does already exist in the table, the launch increments its counter and decrements it once the kernel has terminated. Data entries on the GPU are not replicated. Effectively, the table on the GPU is an overview which parts of the CPU memory are mirrored on the accelerator and how many kernels currently use these mirrored data.

18.4.3   Explicit data transfer

It is the counter semantics that allows us to organise data transfer more explicitly, i.e. outside of a map statement:

Definition 18.10 (*Explicit data transfer steps*) OpenMP allows us to move data *explicitly* to the device, from the device and to update it on the device:

```
#pragma omp target enter data map ...
#pragma omp target exit data map ...
#pragma omp target update data map ...
```

These pragmas are stand-alone pragmas, i.e. they have nothing to do with the block/instruction following them.

While `enter` and `exit` change the GPU's internal reference counter, `update` leaves it unaltered. As a consequence, you can update data from within a target, too. The other two instructions have to be put outside of the compute target region.

As we can separate data movements from computations, it makes sense that there is a `delete` call as counterpart of the allocation. We have `to`, `from`, `tofrom`, `alloc` and `delete`.

Definition 18.11 (*Explicit delete*) There is no need for a `map(delete:x)` clause for the actual compute kernel: The compiler automatically adds the `delete` to any `alloc` or `to` without the corresponding `delete` or `from`. If you however rely on a `map target enter` statement to transfer data, then you have to free the memory explicitly via `from` or `delete`.

Classic kernel structure with explicit data movements

In many codes, core kernels accept an array and write the output to another array, while they accept a few further, often scalar, parameters. They are conceptionally simple. An `enter data map(to:inputData) map(alloc:outputData)` creates the required data structures on the GPU before we launch the kernel. An `exit data map(delete:inputData) map(from:outpuData)` brings the results back and deletes the GPU memory instances again.

It is obvious how the enter and exit annotations are mapped onto GPU table entry modifications. Enter in combination with `to` or `alloc` reserves memory on the accelerator and increments the reference counter in the table. Exit in combination with `from` or `delete` reduces the counter. If you explicitly delete data on the GPU, this delete also requires you to specify the array sizes if the compiler cannot know these at compile time.

It is important to memorise that `alloc` and `delete` mimic the behaviour of `new` and `delete` on the GPU. They expect that a pointer with the same name on the CPU exists before they are called. Our OpenMP philosophy is that OpenMP code remains valid even if a compiler fails to understand our OpenMP pragmas. The data instructions allow us to keep data between different memory regions consistent, but if neither data nor the target blocks are translated, our code remains correct.

Memory errors

When you work with explicit data movements, there are a few evergreens of memory errors:

- If you allocate data on the GPU and you forget the corresponding `from` or `delete`, your GPU will eventually run out of memory. You produce a GPU memory leak.

- If an array on the CPU is changed while there is a replica counterpart on the GPU, the GPU table works with an outdated view of the memory. Never alter a pointer while it is in use on an accelerator.
- If you free memory on the CPU while the corresponding pointer replica on the GPU is still valid, you will eventually get some strange memory error.

18.4.4   Overlapping data transfer and computations

GPU kernels, i.e. `target blocks`, are tasks. The explicit memory transfer operations triggered by `target enter` and `target leave` are tasks, too. Whenever we launch such a GPU kernel on the host, the host blocks (idles) until the corresponding GPU task has finished:

Definition 18.12 (*Target tasks*) All *target blocks are tasks* in OpenMP. However, `target` adds an (implicit) synchronisation (barrier) to its task by default.

With the facts discussed so far, we conclude that

```
#pragma omp target enter data map ...
// call a few very expensive calculations on host
#pragma omp target
{ // the original target but stripped off the map clauses
  // do something on the GPU
}
// we could also do the expensive calculations here
#pragma omp target exit data map ...
// or we could do them here
```

is equivalent to one `target map` where the corresponding data transfer statements are merged into the kernel annotation. Its split-up into explicit data movements and sole compute kernels does not really give us anything, as we work with a *blocking* data transfer model and implicit barriers: We always wait for the GPU to finish its job and until the data transfer is complete before we continue. There are three tasks which are invoked one after the other. Formally, we should model this via four tasks: There's the main task running on the CPU. It launches the data transfer task, waits, launches the compute kernel task, waits, launches another data transfer task, waits, and finally continues.

Like any other task, targets can be augmented with a `nowait` clause. This clause implies that the GPU shall run its computation while the host continues with some other work—maybe meanwhile launching the next kernel. With a `nowait`, we need some further `depends` clauses to ensure that everything on the GPU is executed in the right order (all target instructions also are tasks, so the OpenMP runtime might permute them), but we are now able to make the host computations and the data transfer plus the GPU calculations overlap (Fig. 18.4).

```
#pragma omp target enter data map ... depend(out:x) nowait
#pragma omp target depend(inout:x) nowait
{
    // do something on the GPU
}
#pragma omp target exit data map ... depend(in:x)
#pragma omp task
{
    // do expensive CPU calculations
}
#pragma taskwait
```

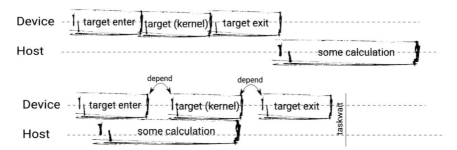

Fig. 18.4 A schematic break-down of the three GPU steps (data transfer, computation, data transfer) followed by some expensive host code on the CPU which is independent of the GPU kernels. With a `nowait`, we can exploit the fact that GPU target invocations are tasks and run the operations all in parallel. Indeed, we "only" use the dependencies and let the runtime decide how and when to schedule a kernel on the GPU

In this code,

- the original data transfer to the GPU is launched as a task. The implicit barrier at the end of the `target` is removed due to the `nowait` clause.
- We immediately launch a second task which is the actual GPU kernel. It does not start before the data transfer task has finished, as it is subject to an in-dependency.
- An additional out-dependency ensures that the task that brings data back is not launched too early.
- Parallel to the three GPU tasks, we spawn a fourth task on the CPU which does all the expensive calculations on the host.
- A final `taskwait` ensures that we wait for both our GPU tasks plus the CPU task. All other (implicit) synchronisation has been removed due to the `nowait` clauses.

x can be an arbitrary placeholder variable. Its content might never be used, it is solely its address that matters. You can distribute the kernels over your whole code, as long as the x remains alive while the kernels on the GPU are still pending or running, respectively.

Multitasking on GPU

Nothing stops you from having multiple (different) tasks on the GPU. Indeed, you can use one data migration task to make multiple compute tasks ready. You can deploy a whole task graph to the accelerator.

18.5 Collectives

GPU tasks can communicate through shared memory and atomics. We know that this allows us to realise all-to-one communication but it remains cumbersome in a GPU context: We have to get all the reduction done on the GPU manually, and then bring the result back to the host.

OpenMP allows us to have reductions over loops. This also works on GPUs. You even can have a reduction over a `distribute` statement, i.e. for multiple teams. That is, reductions also work over multiple SMs. If you use `reduce` in a line which hosts both a `parallel for` and `distribute` clause, then the reduction applies to both clauses. If you split it up, i.e. if you first use `distribute` followed by a separate `parallel for`, then you have to use the same reduction for both clauses to ensure that the data is properly reduced. A reduction implies that the reduced scalar is mapped back from the device to the host.

18.6   Functions and Structs on GPUs

`target` blocks so far are code blocks which work with scalars and arrays over built-in types. It is clear that this gives the OpenMP compiler the opportunity to create bespoke GPU code. It knows all the instructions and exact data layouts.

If you want to use functions within your GPU code, too, you have to tell OpenMP that you need both a host and a device variant. For this, you have to embed the function into

```
#pragma omp declare target
void foo();
#pragma omp end declare target
```

This is a recursive process: If your GPUised function uses another function again, then you have to embed this one into declare target, too.

Definition 18.13 (*Target declarations*) All functions, data structures, variables and classes that you want to use on the GPU have to be embedded into a `declare target` environment.

Declaration versus definition

In most environments I've used so far, you have to embed both the declaration (header files) and the definition (implementation) into declare environments.

Standard libraries for GPUs

It depends quite a lot on your compiler and development ecosystem how you tell OpenMP that you will also need standard library routines on the GPU. For GNU, we usually found

```
-foffload=-lm  -fno-fast-math  -fno-associative-math
```

as linker arguments to do the job.

Structures have to be marked with a `declare`, too, if you want to use them on the accelerator. As soon as the compiler knows that a structure might be used on a GPU, it can analyse its memory layout (that includes its field offsets and arrangement of all the attributes within the structure) and create the counterpart for the accelerator.

If you mark a variable with `declare target`, then this variable is replicated on the GPU at startup. The technique can be used for lookup tables which you want to hold on the accelerator persistently. Finally, you can offload complete classes to a GPU. Just ensure that your declaration wraps both the fields of a struct and all of its routines.

What does not work

A few things do not work or at least are very tricky on GPUs. Some of these items depend on your compiler, so it is a little bit of trial and error:

- You cannot have string routines and outputs to the terminal on the accelerator. Plain `printf` works, but anything that has to do with string construction or concatenation might fail.
- Templates seem to work, but I faced issues multiple times with functors.
- Forget about assertions or similar things that stop your code if they are violated.
- Inheritance, virtual function calls, class attributes (static), and so forth seem to challenge the compiler. Stick to plain classes or, even better, functions.
- We experienced real difficulties with function overloading for some compilers.
- Some of the tests we did with compilers suggest that it is problematic to move classes to and from the accelerator. You might have to map individual (scalar) attributes one by one.

Some important material

- OpenMP Architecture Review Board: *OpenMP Application Programming Interface—Examples*, Version 5.0.0, 2019 https://www.openmp.org/wp-content/uploads/openmp-examples-5.0.0.pdf

> Key points and lessons learned

- CUDA is both a programming language and a chip design "philosophy". One of the core concepts behind GPU architectures is the idea to heavily oversubscribe a core (more threads are running concurrently than what the hardware can handle). The GPU then can swap whole sets of threads (warps) quickly if an instruction has to wait at one point. Every cycle makes progress overall, i.e. no cycles are burnt, if the warp interleaving succeeds. In combination with its enormous hardware parallelism, this idea makes GPUs so fast even though they run at comparably low frequency. Look up Dennard scaling and the energy rule in this context.

- OpenMP supports GPU offloading through the `target` pragma. It launches a new team of tasks and, by default, adds a synchronisation after this.
- We have introduced the term team, league of teams and, in this context, the `distribute` pragma.
- GPUs are particularly efficient if a code relies heavily on classic BSP where the threads barely communicate.
- Data movement in OpenMP can be implicit (managed) or explicit. Explicit data movement pragmas allow us to control the timing of data flows.
- All GPU target regions map onto tasks and we can thus overlap computations on host and GPU as well as data transfer. We can deploy a whole pipeline of steps to the GPU.
- With a declare target, also functions or complex data structures can be outsourced to the accelerator.

Wrap-up

Part V of the manuscript orbits around OpenMP, a pragma-based parallelisation approach in high-performance computing (HPC). Starting from a quick review of machine models and an introduction of the most basic scaling laws, our discussion kicks off with BSP. That is the mindset where OpenMP comes from. We however clarify that modern OpenMP allows us to model task-based parallelism, too, which gives us all the opportunities behind functional decomposition plus data parallelism. In this context, the introduction of GPUs is a step back: Though in principle capable to realise complex parallelisation patterns, GPU hardware is made for massive-scale data parallelism with very homogeneous computation pathways. We may even see this is a step back behind BSP into the realm of SIMD.

OpenMP GPU offloading still is relatively new, i.e. there are still teething problems when we try it out although we may assume that the compilers become more powerful with every new tool generation. It thus makes sense to test your code with different compiler versions and to upgrade the ecosystem frequently. Also, it makes sense to watch the evolution of OpenMP closely, as it is very likely that new features are added to the accelerator part or technical restrictions on the tool side are removed.

Overall, the OpenMP accelerator support remains CPU-centric: It is the CPU which launches (target) tasks, which triggers memory transfers, and, overall, acts as conductor of all computations. This paradigm is not 100% of what many vendors push: GPUs become extremely powerful, host their own interconnects, can send messages to other nodes, and so forth …It might happen that we end up with a situation where the GPUs are the drivers behind an application rather than a CPU. Similar to BSP from where OpenMP's tasking evolved, it might happen that the GPU support eventually will mirror such a paradigm shift, too. It thus might be worth designing new algorithms with a device-centric mindset.

With the basics of numerical algorithms and the toolset how to parallelise on the table, we can realise, more sophisticated ideas how to make codes like our multibody simulation faster and more accurate.

Part VI

Faster and More Accurate Numerical Codes

Higher Order Methods

19

Abstract

To construct higher-order time stepping methods, we discuss two paradigms: On the one hand, we can write down an integral equation for the time stepping and construct more accurate integrators for the right-hand side. On the other hand, we can shoot multiple times into the future to obtain a guess for the additional terms from the Taylor expansion. The first approach leads to multistep methods, the second approach introduces the principles behind Runge-Kutta methods. After a brief discussion how to program arbitrary order Runge-Kutta once the Butcher tableau is known, we briefly sketch leapfrog-type methods which yield particularly elegant, second order methods for our N-body problem.

The explicit Euler takes a snapshot of a system $y_h(t)$ and constructs the next snapshot $y_h(t + h)$ from there through Taylor. If we had two initial values, i.e. $y_h(0)$ and $y_h(h)$, we could make $y_h(2h)$ depend on $y_h(0)$, $y_h(3h)$ depend on $y_h(h)$, and so forth, if we pick a step size of $2h$. There is no formal reason not to do this, though it is clear that the result, even for the simple test equation, now yields some strange solutions with oscillations. We effectively interleave two solvers running with $2h$ each.

© The Author(s), under exclusive license to Springer Nature Switzerland AG 2021
T. Weinzierl, *Principles of Parallel Scientific Computing*, Undergraduate Topics in Computer Science, https://doi.org/10.1007/978-3-030-76194-3_19

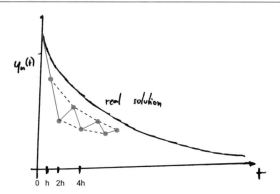

These oscillations fade away for tiny time step sizes, but overall it just does not seem to be a particularly good idea to make two solvers run in a parallel, shifted way.

However, shouldn't we be able to use the history of a solution, i.e. our computational data, to make our predictions better if we have two initial values? Can you sketch such a scheme? Alternatively, could we do multiple shots into the future, and mangle them up into one better guess?

Both questions boil down to the question whether we can get the next terms in the Taylor expansion right. Is there a way to get the $\partial_t \partial_t y(t)$ right even if the ODE is given in first-order formulation? In this case, we do not have to chop it off and ignore it; which makes the $\partial_t^{(3)} y(t)$ dominate the error. We obtain higher accuracy.

Euler belongs into the class of *single step* methods—only the current solution is taken into account—and it is a *single shot* method, as it comes up with the next solution with one F-evaluation.

19.1   An Alternative Interpretation of Time Stepping

In a first attempt to design better time stepping schemes, we reinterpret time stepping as an integration problem. To solve

$$y'(t) = F(t, y(t)),$$

we integrate over time:

$$y(t + h) = y(t) + \int_t^{t+h} F(s, y(s))ds.$$

In practice, this does not get us anywhere, as we do not know the antiderivative of F for interesting problems. This motivated us in the first place to approximate the solution with a computer rather than solving it symbolically. However, we can make

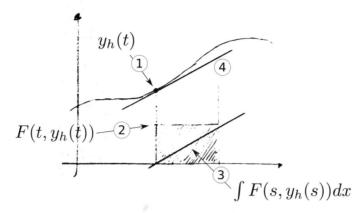

Fig. 19.1 Schematic derivation of the explicit Euler via an integral formulation: (1) We know $y_h(t)$ and thus the value of $F(t, y_h(t))$ (2). If we assume that this value remains constant over the time interval $(t, t + h)$, we can compute the area under $F(s, y_h(s))$. This yields a straight line (3). We can add this straight line to $y_h(t)$ and obtain the value $y_h(t + h)$ (4)

a simple assumption: We can assume that $F(s, y(s)) = F(t, y(t))$ is constant over one time slice. This yields

$$
\begin{aligned}
y(t + h) &= y(t) + \int_t^{t+h} F(s, y(s))ds \\
&= y(t) + \int_t^{t+h} F(t, y(t))ds \\
&= y(t) + F(t, y(t)) \cdot \int_t^{t+h} 1\,ds \qquad (19.1) \\
&= y(t) + F(t, y(t)) \cdot s|_{s=t}^{t+h} \\
&= y(t) + F(t, y(t)) \cdot h.
\end{aligned}
$$

Section 12.3 derives the explicit Euler through a Taylor series. The above sketch yields exactly the same algorithm with integration glasses on (Fig. 19.1). It is clear that both approaches produce the same formula in the end—otherwise, one formalism would be wrong. On the following pages, we exploit this equivalence to construct so-called multistep algorithms:

19.2 Adams-Bashforth

For the explicit Euler, we assume that the time step $y_h(t)$ gives us an $F(t, y_h(t))$ which then does not change until we hit $y_h(t + h)$. Equation (19.1) formalises that. Let's assume that we know one further previous time step $y_h(t - h)$, too. We therefore also know the value of F at $t - h$. In the thought experiment with step size 2h,

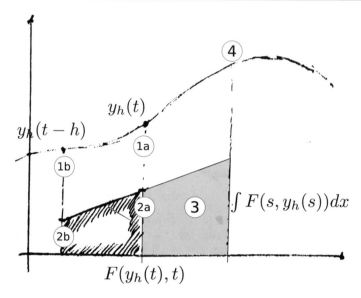

Fig. 19.2 Schematic derivation of Adams-Bashforth using notation from Fig. 19.1: This time, we assume we know both $y_h(t)$ (1a) and $y_h(t-h)$ (1b). Therefore, we also know $F(t, y_h(t))$ (2a) and $F(t, y_h(t))$ (2b). We take these two data points and extrapolate what F looks like over the interval $(t, t+h)$. This yields an area under a straight line (3). We integrate over this area to obtain a new value $y_h(t+h)$ (4)

we know two derivatives in the point y(h): We have the derivative from y(0) and we have the derivative at y(h). Instead of following either of them, and instead of switching between different solutions, we try to combine these two pieces of information cleverly.

With additional information from previous time steps, we can make a more educated guess for the integrand (Fig. 19.2): Let F not be constant anymore, but assume it grows linearly! As we know $F(t-h, y_h(t-h))$ and $F(t, y_h(t))$, we draw a line through both values and assume that this F continues linearly beyond t until we hit the next snapshot at $t+h$.

Definition 19.1 (*Multistep methods*) *Multistep methods* keep track of previous solutions to come up with a more educated guess for F. They use historical information to predict what F looks like over the next time span.

We redo the (symbolic) integration over our improved approximation of F and obtain a new formula where the new time step depends on $y_h(t)$ as starting point (same as Euler), the value of F around that starting point (same as Euler) plus an additional term which depends on the difference of F between t and $t-h$ (not the same as in the Euler construction):

$$y_h(t+h) = y_h(t) + \int_t^{t+h} F(s, y(s))ds$$

$$= y_h(t) + \underbrace{F(t, y_h(t)) \cdot h}_{\text{Area under constant function}}$$

$$+ \underbrace{\underbrace{\overbrace{\frac{F(t, y_h(t)) - F(t-h, y_h(t-h))}{h}}^{\text{Slope of better approximation}} \cdot h}_{\text{Walk } h \text{ along slope}} \cdot \frac{h}{2}}_{\text{Area under triangle}}$$

$$= y_h(t) + \int_t^{t+h} F(s, y(s))ds$$

$$= y_h(t) + h \cdot (\frac{3}{2}F(t, y_h(t)) - \frac{1}{2}F(t-h, y_h(t-h))).$$

Definition 19.2 (*Adams-Bashforth two-step method*) The *Adams-Bashforth* two-step method is given by

$$y_h(t+h) = y_h(t) + h \cdot (\frac{3}{2}F(t, y_h(t)) - \frac{1}{2}F(t-h, y_h(t-h))).$$

The multistep methods here do not really care about old solutions before our last time step $y_h(t)$. They care about the F values at previous time steps! There are three important implications for programmers:

1. We have to bookkeep F values from previous time steps. This will increase our memory footprint.
2. We make certain assumptions about the smoothness of F in time. We take it for granted that the history of F tells us how F develops in the future.
3. We need historic data when we kick off the solve. When we solve an IVP, we typically do not have such historic data. In this case, we can for example run explicit Euler steps (with tiny time step sizes) until we have a sufficient set of historical F data to switch to our multistep method of choice.

There are more sophisticated multistep methods than Adams-Bashforth. We omit a discussion of these. A diffent type of higher order methods is more popular in science:

19.3 Runge-Kutta Methods

Adams-Bashforth uses historical information to construct better estimates of y_h. If we don't want to look back and use historic data, we can do the opposite thing: Shoot

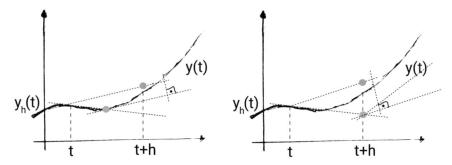

Fig. 19.3 Schematic illustration of Runge (RK2; left) and Heun's method (right): RK2 runs an explicit Euler, but only for $h/2$ (end up in orange circle). It then evaluates the F value (gradient) of the ODE there and uses this F value to kick off a whole "Euler" step (end up in green circle). Heun first runs an explicit Euler for the whole h interval (orange circle). It then takes half of the F value at this predicted point plus half of the original F value (gradient) to perform one time step from the starting point (end up in green circle)

into the future multiple times with different time step sizes h and thus obtain a feeling how F changes over time. Such schemes are also called *multiple shooting methods*.

Here is a sketch of a very simple multiple shooting method: With the explicit Euler, we use the F value at time t and assume that it is constant over the whole interval $(t, t + h)$. If we want to stick to that concept that F is constant, we could evaluate $F(t + h/2)$ and use this one as a better approximation to F. Unfortunately, we don't know $F(t + h/2)$. Therefore, we use two shots:

- We run an explicit Euler step for half of the time step to get an estimate for a $\hat{y}(t + h/2)$. The hat emphasises that this is a helper result that we do not use directly in the end.
- We plug this estimate into the right-hand side of the ODE, i.e. compute $F(t + h/2, \hat{y}(t + h/2))$.
- We finally return to $y_h(t)$ and use this second F-value to realise an Euler step over the whole time interval.

Schemes as the one sketched are also called predictor-corrector schemes (Fig. 19.3): We compute an estimate of F and feed it into the time step. We know that it is a guess, i.e. we can call it a prediction of what the right-hand side looks like. Next, we use this prediction to compute a new solution which we use to replace or correct the original guess. In the introductory example, we walk for 2h along the derivative, but stop at h to construct a second derivative. We can use these two directions to construct two independent solves. Or we can combine the two derivatives into one, better guess and use this one instead of either of them.

Definition 19.3 (*Runge's method*) *Runge's method* is given by

$$y_h(t + h) = y_h(t) + h \cdot F(t + h/2, y_h(t) + h/2 \cdot F(t, y_h(t))).$$

Runge's method is more accurate than the explicit Euler, and yet it does not use any historic data. It does not make the assumptions that the history of F tells us something about its future. It makes an attempt of what y_h would look like half a time step down the line, and then evaluates the F there again. No technical reason forces us to do half a time step for the first guess:

Definition 19.4 (*Heun's method*) For *Heun's method*, we also use two shots, but the first attempt spans the whole h. This attempt (prediction) yields a new guess for F. We average the F at the current time step and the predicted F:

$$y_h(t + h) = y_h(t) + h \left(\underbrace{\frac{1}{2} F(t, y_h(t))}_{\text{Euler part}} + \underbrace{\frac{1}{2} F(t + h, y_h(t) + h F(t, y_h(t)))}_{\text{Test shot}} \right).$$

Whenever we see such a "let's shoot into the future" algorithm as computer scientists, we can immediately generalise it via recursion:

• Compute $k_1 = F(t, y_h(t))$;
• compute new shots $s \geq 2$ via

$$k_s = F(t + c_s h, y_h(t) + c_s h \cdot \hat{F})$$

where the $\hat{F}$ results from a weighted linear combination of all previous k_s;
• combine the s shots via

$$y_h(t + h) = y_h(t) + h (b_1 k_1 + b_2 k_2 + \ldots). \tag{19.2}$$

For an *s-step multiple shooting method*, we need to compute s attempts (predictions). Each of these enters the final sum subject to a weight b_s. To compute a new shot, we march always a fraction $c_s \leq 1$ of the whole time span and combine the previous estimates. As we replace the $c_s h \cdot \hat{F}$ from above with a linear combination of previous shots we end up with

$$k_s = F(t + c_s h, y_h(t) + h \sum_{i=1}^{s-1} a_{si} k_i), \tag{19.3}$$

where a $c_1 = 0$ kicks off the recursive construction of the ks.

There are plenty of weights in this description. For two steps, the selection $b_1 = 0$, $b_2 = 1$ and $c_2 = a_{21} = 0.5$ gives us Runge's method. The combination $b_1 = 0.5$, $b_2 = 0.5$, $c_1 = a_{21} = 1$ yields Heun's method. Before we sketch how these parameters are derived and make further statements on the order of the methods, we introduce a compact way to present the weights entering the equations.

Systems of ODEs

If you are dealing with systems of ODEs, you will have to develop the scheme in all unknowns, i.e. the ks are time derivatives of all of your unknowns. If you have, for example, N particles and each particle has three position and velocity components, your ODE system has $2 \cdot 3 \cdot N$ entries and each k therefore has $6N$ entries, too (cmp. discussion on page 142).

That is, for each trial shot, you have to compute new trial positions plus velocities of all particles using your F approximations, i.e. the velocity and position updates computed so far; subject to weights a_{si}. Once you have these trial properties, you compute the new forces plus the corresponding position updates. The new forces plus positions then feed into the final position and velocity updates. All intermediate vectors have to be stored.

Definition 19.5 (*Butcher tableau*) The weights behind Runge-Kutta methods are something you typically look up. The table where you look them up is called *Butcher tableau*.

A Butcher tableu is written as

$$
\begin{array}{c|cccc}
0 \\
c_2 & a_{21} \\
c_3 & a_{31} & a_{32} \\
c_4 & a_{41} & a_{42} & a_{43} \\
\hline
 & b_1 & b_2 & b_3 & b_4
\end{array}
$$

This tableau formalises a four-step method. We read it line by line. The value $k_1 = F(t, y_h(t))$ is our starting point. We always need this one. From hereon, we compute $k_2 = F(t+c_2 h, y_h(t)+h \cdot a_{21} k_1)$. This formula uses the indices from the second line from the tableau. The terms k_3 and k_4 are then computed accordingly. Once we have all four approximations of F, we use a linear combination to construct our "real" F. The weights b_i for this linear combination are given in the tableau's bottom line.

Runge's method is fully specified by

$$
\begin{array}{c|cc}
0 \\
\frac{1}{2} & \frac{1}{2} \\
\hline
 & 0 & 1
\end{array}
$$

An explicit Euler corresponds to the degenerated tableau

$$
\begin{array}{c|c}
0 \\
\hline
 & 1
\end{array}
$$

Heun's method finally is given by

$$
\begin{array}{c|cc}
0 \\
1 & 1 \\
\hline
 & \frac{1}{2} & \frac{1}{2}
\end{array}
$$

The most popular Runge-Kutta variant in papers is the fourth-order Runge-Kutta method.

$$
\begin{array}{c|cccc}
0 \\
\frac{1}{2} & \frac{1}{2} \\
\frac{1}{2} & 0 & \frac{1}{2} \\
1 & 0 & 0 & 1 \\
\hline
 & \frac{1}{6} & \frac{2}{6} & \frac{2}{6} & \frac{1}{6}
\end{array}
$$

Runge-Kutta notation

When authors refer to Runge-Kutta without specifying the number of steps, they typically mean four steps. Sometimes, people write RK(4).

19.3.1   Accuracy

For us, copying indices from the Butcher tableau into our code is usually sufficient. Yet, one should have an intuition where the parameters in the tableau come from. For this, we return to the Taylor series—and use it multiple times. Assume we want to write down a scheme which is second order accurate. We want to realise

$$y_h(t+h) = \underbrace{y_h(t)}_{\text{given}} + \underbrace{\frac{h}{1!}\partial_t y_h(t) + \frac{h^2}{2!}\partial_t^{(2)} y_h(t)}_{\text{we want to get this right}} + \underbrace{\cdots}_{\text{acceptable error}} .$$

The explicit Euler only approximates $\frac{h}{1!}\partial_t y_h(t) = hF(t, y_h(t))$ correctly. This is our predictor step. The idea behind a corrector is that the next step handles the remaining $\frac{h^2}{2!}\partial_t^{(2)} y_h(t)$. While I usually use ∂_t and d_t quite sloppily, I now carefully distinguish them, as $F(t, y_h(t))$ depends on t in both arguments.

Eliminate second derivative

With the chain rule and basic calculus, the second derivative can be rewritten as

$$\partial_t (\partial_t y_h(t)) = d_t F(t, y_h(t)) \quad \text{(watch out for symbol switch)}$$
$$= \partial_t F(t, y_h(t)) + \partial_{y_h(t)} F(t, y_h(t)) \cdot \underbrace{\partial_t y_h(t)}_{=F(t, y_h(t))} . \tag{19.4}$$

- We replace the $\partial_t y_h(t)$ by the ODE definition.
- As $F(t, y_h(t))$ depends on t both directly (first argument) and indirectly (second argument), we have to use the total derivative d_t after the replacement.
- We derive w.r.t. t first (that's a direct t-dependency) and then w.r.t. the second parameter. The second one is a function depending on t, so we have to apply the chain rule.
- Finally, we replace the $\partial_t y_h(t)$ with the ODE once again.

The Taylor expansion with a second-order term contains terms with $F(t, y)$, with $\partial_t F(t, y)$ and $\partial_y F(t, y)$. All second derivatives are eliminated.

We next realise the following strategy: Write down the definition of the s-step method and unroll the recursive expressions therein. We continue to demonstrate this for the $s = 2$-step method:

$$y_h(t+h) = y_h(t) + h(b_1 k_1 + b_2 k_2) \quad \text{from (19.2) which unfolds into}$$
$$= y_h(t) + hb_1 \cdot F(t, y_h(t)) + hb_2 \cdot F(t + c_2 h, y_h(t) + ha_{21} \cdot F(t, y_h(t))).$$

The next step is the clue: The unrolled recursive formula expects us to evaluate F at a time $t + c_2h$. It is time for Taylor!

$$F(t + c_2h, \ldots) = F(t, \ldots) + c_2h\partial_t F(t, \ldots) \quad \text{respectively}$$

$$F(\ldots, y_h(t) + ha_{21} \cdot F(t, y_h(t))) = F(\ldots, y_h(t)) + ha_{21} \cdot F(t, y_h(t))$$
$$\partial_{y_h(t)} F(\ldots, y_h(t)). \tag{19.5}$$

The last trick reduces all terms to rely on F evaluations at (t, y_h). The terms in the generic Runge-Kutta formulation (19.2) thus match the terms in the Taylor expansion. In the $s = 2$-example,

$$
\begin{aligned}
y_h(t + h) &= y_h(t) + hb_1 \cdot F(t, y_h(t)) + hb_2 \cdot F(t + c_2h, y_h(t) + ha_{21} \cdot F(t, y_h(t))) \\
&= y_h(t) + hb_1 \cdot F(t, y_h(t)) + hb_2 \cdot F(t, y_h(t)) \\
&\quad + h^2 b_2 c_2 \cdot \partial_t F(t, y_h(t)) \\
&\quad + h^2 b_2 a_{21} \cdot F(t, y_h(t)) \cdot \partial_{y_h(t)} F(t, y_h(t)).
\end{aligned}
$$

This is a plain "plug our equations in" exercise. The only pitfall is that the Taylor expansions in (19.5) run either in the direction of the first argument or the second.

As the term structure in (19.5) and above match now, we can pick the free weights such that the equations are exactly the same. However, (19.3) gives us four unknowns to play around. The fact that there are four "free" parameters (b_1, b_2, c_2 and a_{21}) explains why there are (infinitely) many Runge-Kutta schemes per order. We have some freedom of choice.

In the example above, we collect all the h terms and obtain the constraint $b_1 + b_2 = 1$. We furthermore collect the h^2 terms and obtain $b_2 c_2 = \frac{1}{2}$ as well as $b_2 a_{21} = \frac{1}{2}$. Both Runge's method and Heun's method match these constraints. We conclude that they both yield the same order of accuracy.

The exercise above constructs a second-order scheme. All techniques can be applied recursively, i.e. over and over again, to construct schemes with order $p > 2$. Yet, there are bad news for RK methods: You do not always get a pth order accurate integrator after s steps. With proper parameter choices, we can get away with $p = s$ for $p \in \{3, 4\}$. To construct higher-order RK methods ($p > 4$), we however will need more than s steps to get enough free parameters for tailoring them such that the result matches the higher order Taylor expansion terms. Furthermore, finding these weights becomes nasty and the number of conditions which your parameters in the Butcher tableau have to meet explodes.

Definition 19.6 (*Butcher barrier*) For $p \geq 5$, there are no generic, explicit Runge-Kutta construction rules with $s = p$. $p \geq 5$ is the *Butcher barrier*.

19.3.2 Cost and Parallelisation

While RK(2) (Runge's method) or Heun's method both are more accurate than Euler, we have a price to pay: We need two F evaluations per time step rather than one. Let us assume that both a second-order scheme and an explicit Euler yield a result of roughly the same quality for a particular h. To reduce the error of a calculation by a factor of four with standard explicit Euler, we have to divide the time step size by four. We have to half it twice. Instead of computing with $h/4$, we could use only $h/2$ with RK(2). As it is second order globally, one refinement step in h reduces the error by a factor of four, too. However, each RK(2) step requires us to evaluate F twice. If we assume that F is by far the most expensive part of our calculation, we run into a tie: The more accurate method needs half the steps, but each step has twice the cost. Along these arguments, things change if we switch to RK(3) or RK(4).

While RK(3) or RK(4) exhibit an advantageous cost-to-accuracy ratio, other drawbacks cannot be easily compensated for:

- We have to bookkeep the predictions k_s. They are intermediate results that feed into the subsequent predictions, but they also contribute to the final linear combination. Therefore, we have to store them throughout the computation. More shots means bigger memory footprint.
- The rightmost entries in every line of the Butcher tableau are unequal to 0. That is, the evaluation of k_s requires the completion of all computations for k_{s-1}. The "outer loop" running through the shots does not exhibit trivial concurrency. Instead, each s-step introduces a parallel barrier.
- In a distributed memory world, each s-step requires a complete data exchange. If we have N particles, we have to compute k_s for every single particle before we compute the k_{s+1} for all particles in the next sweep. k_{s+1} will need an updated guess of the particle properties (positions).

The three reasons above make many supercomputing codes ignore higher-order RK methods. Instead, there is one particularly popular scheme which is second order (almost) for free:

19.4 Leapfrog

When we write code for our second-order ODEs describing particle motion—where the acceleration is affected by the particles' position—we first rewrite these ODEs into a system of first-order quantities (cmp. discussion on page 141). Pseudo code for the resulting scheme consists of a sequence of three loops per time step:

1. Compute the force F (or acceleration) acting on each particle.
2. Update the position of each particle.
3. Update the velocity of each particle.

You can fuse the second and the third loop to get rid of one loop. This results directly from the respective control flow graph. If we permute loop two and three, we update the velocity first and this new velocity enters the position immediately. This modification of the explicit Euler is a different algorithm and is called *Euler-Cromer*.

Invalid optimisation

I have seen multiple "optimised" codes which try to fuse the force computation loop with the updates. The motivation behind such a merger is valid: Update loops are computationally cheap and thus hard to scale up. If we can hide them behind another loop, we get code that scales better.

Unfortunately, such optimisations are vulnerable to bugs: We must not compute the forces on a particle, update the particle's positions plus velocity, and then continue with the next particle. We have to ensure that one, invariant particle configuration feeds into all calculations of the follow-up step. If we change some particles in the preimage on-the-fly, we end up in a situation where some particles are already a step ahead and feed into calculations, while others have not been updated yet. We get inconsistent data.

A totally new scheme arises, once we decide not to stick to integer time steps anymore: Leapfrog time stepping keeps the position discretisation, but shifts the execution of the velocity updates (Fig. 19.4): The velocities are not updated per time step but after half a time step, after 1.5 time steps, and so forth. Velocity and position updates permanently overtake each other, i.e. jump over each other.

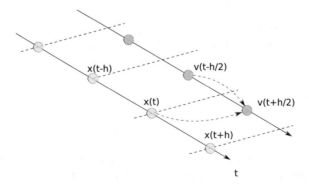

Fig. 19.4 The idea behind leapfrog is that velocity and position updates happen in the same time intervals but are set off against each other by $h/2$. The dotted arrows indicate which snapshots influence $v(t + h/2)$. Intuitively, it is clear why leapfrog is more accurate than Euler: The value of $v(t + h/2)$ requires data from $v(t - h/2)$ and (formally) from $x(t + h/2)$. The latter is not available (yet) and we thus extrapolate it from $x(t)$ using the knowledge how it will evolve. This extrapolation/knowledge argument is the reason behind the improved accuracy

Definition 19.7 (*Leapfrog time stepping*) The blueprint of *leapfrog time stepping* is given by

$$x_h(t + h) = x_h(t) + h \cdot v_h(t + h/2), \tag{19.6}$$

and

$$v_h(t + h/2) = v_h(t - h/2) + h \cdot F(t, x_h(t)).$$

To analyse the accuracy of this scheme, we have to map it back onto a Taylor expansion. The formal problem with the position update in (19.6) is that it has this fractional time step in there, and such a time step does not arise in Taylor. The same argument holds for the velocities. The dilemma is resolved by recognising that the explicit Euler extrapolates linearly and that it works both forward and backward in time:

Time-reversibility

The Taylor expansion works forward and backward in time, i.e. you can develop $f(t + h)$ around t through the Taylor series as well as $f(t - h)$. Most notably, the derivatives entering the Taylor expansions are exactly the same in

$$f(t + h) = f(t) + h \cdot \partial f(t) \quad \text{and} \quad f(t - h) = f(t) - h \cdot \partial f(t).$$

The expansion is *time-reversible*.

The explicit Euler is not time-reversible. It yields a concatenation of straight lines (Fig. 13.1), and we thus have the relations

$$f(t + h) = f(t) + h \cdot \partial_t f(t) \quad \text{and} \quad f(t - h) = f(t) - h \cdot \partial f(t - h).$$

Our explicit Euler does not preserve the time-reversibility.

Numerics philosophy

Phenomena in nature have plenty of properties (energy conservation, time reversibility, smoothness, …), which we might loose when we solve them on a computer. If a simple trick (such as the shift of the operator evaluations) helps us to regain some of these properties, we typically get more robust code plus new opportunities to design better numerical schemes; that allow us to pick larger time step sizes preserving the same accuracy, e.g.

The leapfrog is different. If we study the solution around $\{t, t + h, t + 2h, \ldots\}$, the solution is time-reversible in the velocity. If we study the solution around $\{t + h/2, t + 3h/h, t + 5h/2, \ldots\}$, the solution is time-reversible in the position. We can therefore rewrite (19.6) into

$$x_h(t + h) = x_h(t) + h \left(\underbrace{v_h(t - h/2)}_{\text{go back in time}} + h \cdot F(t, x_h(t)) \right)$$

$$= x_h(t) + h \Big(v_h(t) - h/2 \cdot F(t, x_h(t)) + h \cdot F(t, x_h(t)) \Big)$$

$$= x_h(t) + h \cdot \underbrace{v_h(t)}_{=\partial_t x_h(t)} + h^2/2 \cdot \underbrace{F(t, x_h(t))}_{=\partial_t \partial_t x_h(t)}.$$

This is exactly the Euler expansion where we truncate after the third term. As the same train of thought holds for the velocities, too, the leapfrog is globally second-order accurate.

Velocity Verlet

In the context of particle methods, we often use *Velocity Verlet* time integrators. They are alternative formulations of leapfrog, which yield velocity values at the same points in time where we also encounter position updates. Having the velocity explicitly at hand at these points is important for visualisation and a lot of postprocessing or coupling to other models. Velocity Verlet relies—in its simplest form—on the following steps:

1. Assume that $x_h(t = 0)$ and $v_h(t = 0)$ are given as initial values. $t = 0$ initially.
2. Compute $v_h(t + h/2) = v_h(t) + h/2 \cdot F(t, x_h(t))$.
3. Compute $x_h(t + h) = x_h(t) + h \cdot v_h(t + h/2)$ from (19.6).
4. Compute $v_h(t + h) = v_h(t + h/2) + h/2 \cdot F(t + h, x_h(t + h))$.
5. Continue with step 2 for $t \leftarrow t + h$.

After step 4, both position and velocity are valid for this point.

Some important material

- P. Deuflhard, *Recent Progress in Extrapolation Methods for Ordinary Differential Equations*. SIAM Review, 27(4), pp. 505–535 (1985) https://doi.org/10.1137/1027140
- D. Ketcheson, U. bin Waheed, *A comparison of high-order explicit Runge–Kutta, extrapolation, and deferred correction methods in serial and parallel*. Communications in Applied Mathematics and Computational Science, 9(2) (2014), pp. 175–200 https://arxiv.org/abs/1305.6165
- W. Gander, M.J. Gander, F. Kwok, *Scientific Computing—An Introduction using Maple and MATLAB*. Springer (2014)
- A. Cromer, *Stable solutions using the Euler Approximation*. American Journal of Physics, 49 (455) (1981) https://doi.org/10.1119/1.12478

> **Key points and lessons learned**

- Multistep methods keep track of some previous time steps and construct a smooth curve for F through these steps to predict the real F over the upcoming time interval.
- Runge-Kutta (RK) methods are the most prominent type of multiple shooting methods. The principle here is that we shoot into the future using all known approximations of F so far, and then take the result to obtain a new F guess. This works recursively.
- Butcher tableaus provide us with a formal specification how to program a Runge-Kutta (RK) method.
- Leapfrog-type methods are very popular with particle methods where we maintain velocity plus position anyway.

Adaptive Time Stepping

20

Abstract

The discussion around A-stability makes it clear that ODE integrators are very sensitive w.r.t. time step choices. Some simple thought experiments illustrate that a fixed time step size for a whole simulation often is inappropriate. We sketch one simple multiscale approach to guide h-adaptivity (adaptive time step choices) and one simple p-adaptivity strategy.

Assume you have to create a bike map of two tracks. More specifically, you want to create a height profile, i.e. how these tracks go up and down. There are two different tracks on your todo list: One track is a single track mountain bike trail with a lot of bumps. The other trail is a dismantled railway line.

If you ride along a former train route, you don't need to take many snapshots per time span to track the train track's altitude. Railway lines are smooth and do not ascend or descend suddenly. No need to invest a lot of snapshots. Things

© The Author(s), under exclusive license to Springer Nature Switzerland AG 2021
T. Weinzierl, *Principles of Parallel Scientific Computing*, Undergraduate Topics
in Computer Science, https://doi.org/10.1007/978-3-030-76194-3_20

are different on the mountain bike trail: Here, you frequently have to make snapshots to track all the features of the ground.

While a certain snapshotting strategy might work for one particular trail type, mixed-type trails require mixed strategies. Assume you bike up a former railway line to ride downhill across the fields next to it. To map such a circular loop, a low snapshot frequency is economically reasonable for the uphill path, but the downhill ride still requires a lot of snapshots. Having one fixed snapshotting strategy is not a good idea.

In Sect. 13.1, we encounter a situation where a time integrator yields oscillations: The test equation with a very steep gradient gives you meaningless results if it is subject to too large time step sizes and an explicit Euler. Yet, it is not the gradient of the solution that causes issues. It is the massive change in the gradient's magnitude. If the analytical solution were a straight line with a high derivative, explicit Euler would be perfectly accurate and stable.

The Dahlquist test equation illustrates the problem we face with massive gradient changes, and it also allows us to formalise the problem that we face: When we introduce the term A-stability on page 153, we observe that we can always find, for a fixed h, an α in $y(t) = e^{-\alpha t}$ such that the explicit Euler yields instabilities.

Definition 20.1 (*Stiff problem*) A *stiff problem* is a problem that requires conditionally stable algorithms to use extremely small time step sizes.

If codes have to use tiny time step sizes, we have to recognise this. If they do not need tiny time step sizes all the time, we should come up with an algorithm that uses tiny step sizes where required but maximises h otherwise. We want to balance approximation errors versus compute effort.

20.1 Motivation: A Very Simple Two-Grid Method

We simulate a billiard table with an explicit Euler. It is clear that it is almost impossible to make the billiard balls totally rigid and incompressible: As we cut time into slices, i.e. we study snapshots, there will always be setups where a ball should have bumped into the other one and pushed it to the side, but the simulation made it penetrate it. A simple fix wraps the balls into rubber cushion. If two balls move away from each other or if they are further away than a hard-coded ϵ, we just simulate how they roll on the table. If they approach each other and the distance d between them becomes $d \leq 2\epsilon$, we accelerate them with $(2\epsilon - d)^{12}$ to push them away from each other. With an exponent of 12, the push force "explodes" as the particles approach. This explosion approximates a real rigid body collision where the force is infinite for penetrating objects. We call this is a soft collision model (Fig. 20.1).

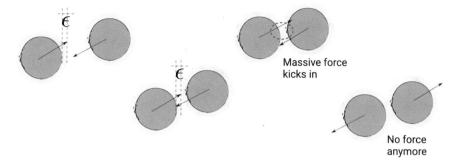

Fig. 20.1 Left to right: Spherical rigid bodies are surrounded by a rubber halo of size ϵ which is compressible. If they approach each other, there's no force in-between them as long as they are more than 2ϵ away from each other. Once their halos overlap, a massive force is applied to push them away from each other. If they move away from each other, there is no force interaction (anymore)

For very fast billiard balls, the soft collision model continues to fail. We add forces once two balls are closer than 2ϵ, but we still assume that they never really penetrate. If they roll very fast, we therefore still need tiny time step sizes.

It is not a trivial exercise to find out whether and when we need small sizes. One solution is to design an algorithm which always runs the Euler with two time step sizes h and $h/2$ concurrently. If the $h/2$ setup yields reasonably similar results to the h-simulation, we accept the time step with h and continue. If not, we rerun our calculation with $h/2$ and $h/4$, i.e. we half the hs.

Some people call this a *two-level, feature-based refinement criterion*. Let's use the scalpel: It takes two "levels"—one with h and one with $h/2$—and compares the outcomes of the two runs. The $h/2$ then acts as a guardian for the real simulation with h. It compares the features of the solution. We don't know the real solution to our simulation. Yet, if there are large differences between two solutions running with h versus $h/2$, we better be careful and use small time step sizes.

Coarsening

The text sketches only a refinement criterion, as we discuss when to refine h. For a full-blown code, we would likely complement this refinement criterion with a time step coarsening criterion making h larger again if this is admissible.

20.2 Formalisation

We want to solve our problem as quickly as possible. At the same time, the error should be bounded. Both goals can only be achieved by ensuring that all time steps produce an error of roughly the same size: If there is one single time step producing a massive error, we know that this massive error propagates in time and eventually destroys the quality of our final outcome. In this chapter, we distinguish two different algorithmic strategies how to reduce the compute cost, while we keep the error under control:

1. Make the time step size h as big as possible. Large time step sizes imply that we need fewer steps to simulate up to a given time stamp.
2. Pick a numerical scheme per time step which is as cheap as possible. As discussed throughout Chap. 19, this often means switching to higher-order methods where appropriate, i.e. where the solution is smooth.

Obviously, both strategies can be combined.

Definition 20.2 (*h- vs. p-adaptivity*) If we change the time step size, we call this *h-adaptivity*. If we change the (polynomial) order of our scheme, we call this *p-adaptivity*.

Both strategies rely on an error estimator. As we don't know the real solution, we don't know the error of our numerical approximation either. However, we can try to construct a (pessimistic) guess of what the error looks like:

Definition 20.3 (*Error estimator*) Let $e_h(t)$ describe the local error of a numerical scheme. A function $\tau(t)$ with $|e_h(t)| \leq \tau(t)$ is an *error estimator*.

The idea behind an error estimator is that we use τ to determine which numerical model to use and which time step size. With the most aggressive choice, we still ensure $\tau(t) \leq tol$ per time step, with a fixed global tolerance tol. We distinguish two different selection strategies:

Definition 20.4 (*Optimistic vs. pessimistic time stepping*) An *optimistic time stepping* approach constructs an estimate $y_h(t + h)$ and then evaluates $\tau(t + h)$. If $\tau(t + h) \leq tol$, it accepts the solution, i.e. continues with the time stepping. Otherwise, is performs a *rollback*, i.e. returns to $y_h(t)$ and runs the time step again with a modified numerical scheme/parameters. A *pessimistic* implementation always chooses a stable numerical scheme right from the start.

As we do not know the correct solution, pessimistic implementations usually work with large security thresholds. They always damp the time step size significantly, to ensure that they stay on the safe side. They have to be "too careful" with the parameter choices. Optimistic implementations do not face this problem, but they require the programmer to backup solutions such that the code can roll back upon demand. While they promise to yield a better time-to-solution as they need no excessive security distance, they can become slower than their pessimistic counterpart if they have to roll back too often or if the backups become too expensive.

   A classic example of a hybrid is a rolling average strategy: I take a time step size $h^{(guess)}$ for the time step. After the time step, I compute $\tau(t + h^{(guess)})$ to derive a $h^{(adm)}$, i.e. the largest admissible time step size that would have worked. If we have been on the safe side, i.e. chosen a too small time step size, I use the average between $h^{(guess)}$ and $h^{(adm)}$ as next time step size. I creep towards the admissible time step size. Otherwise, I roll back and restart the time stepping with $h^{(adm)}$.

20.3 Error Estimators and Adaptivity Strategies

20.3.1 Adaptive Time Stepping

Let's exploit the fact that we know the local convergence order of our time stepping scheme. As we truncate the Taylor series,

$$\tau_h(t + h) = Ch^p. \tag{20.31}$$

Although we do not know C, we have h (the time step size) and we know p from our numerical analysis. The remainder is a calibration problem. We have to determine C from two measurements:

If we rerun our simulation with half of the time step size, we get $\tau_{h/2}(t + h/2)$ and, after one more time step a $\tau_{h/2}(t + h)$. The definition of the two τ terms yields

$$y(t + h) = y_h(t + h) + \tau_h(t + h) \quad \text{and} \tag{20.32}$$
$$y(t + h) \approx y_{h/2}(t + h) + 2\tau_{h/2}(t + h).$$

In the second line of (20.32), we assume that the local truncation error $\tau_{h/2}$ is roughly the same over both time steps, i.e. local $\tau_{h/2}(t + h/2) \approx \tau_{h/2}(t + h)$. Indeed, if the solution is reasonably smooth, the error behind this assumption is negligible. Next, we insert what we know about the shape of τ, i.e. we exploit that

$$\tau_{h/2}(t + h) = C \left(\frac{h}{2} \right)^p = \frac{1}{2^p} \tau_h(t + h).$$

This expansion assumes that the C is the same for both runs. We are only interested in one versus two time steps. Substracting the two equations (20.32) from each other gives us

$$y_{h/2}(t + h) - y_h(t + h) = (1 - \frac{2}{2^p})\Big(\tau_h(t + h)\Big), \quad \text{i.e.}$$

$$|\tau_h(t + h)| \leq \frac{1}{1 - \frac{2}{2^p}} |\left(y_h - y_{h/2} \right)(t + h)|.$$

With two measurements with two different h choices, we can compute which h is admissible. For this, we insert our ansatz $\tau_h(t + h) = Ch^p$ and set the absolute value of this expression equal to tol.

$$C \approx \frac{1}{1 - \frac{2}{2^p}} |y_h(t + h) - y_{h/2}(t + h)| h^{-p} \quad \text{and therefore}$$

$$C \left(h^{\text{(adm)}} \right)^p \leq tol \Leftrightarrow h^{\text{(adm)}} \leq \sqrt[p]{\frac{tol}{C}}. \tag{20.33}$$

With only three time steps, we get a formula how to determine an optimal admissible time step size for our setup, i.e. a maximal time step size for a given tolerance. This is a check we can do in-situ, i.e. while our code is running. In practice, there are a few challenges to keep in mind:

- The approach works only as the ansatz (20.31) defines a function that runs through 0. In theory, it is correct that τ runs through zero—otherwise, our scheme would not be consistent. In practice, it does not run through zero. Once h becomes too small, we enter a regime where the round-off errors of floating point arithmetics dominate the result.
- The ansatz (20.31) works with a constant p where we know p a priori and observe this order in our data. It is important to remind ourselves that we always use a fundamental assumption when we construct integration schemes: For a numerical scheme to have pth order locally, the analytical solution has to exhibit this smoothness, too. If the smoothness of our solution changes (suddenly), the error estimator breaks down, too. This reminder is of limited relevance for Euler (we can't work "simpler" than that), but it plays a big role for higher-order schemes.
- The approach works as long as the ODE's right-hand side F is well-behaved and does not change dramatically over the time step under examination. If the F suddenly becomes very large in the second smallish time step, then the τ is not equally distributed among the two intervals with $h/2$ anymore.

20.3.2 Local Time Stepping

Many interesting phenomena in science span multiple (time) scales. Our planets orbit around the sun, but this is a very slow process compared to their moons which tend to spin around their respective planets. If we want to simulate the solar system, the moons make the numerical treatment cumbersome: As they are so light and, hence, fast and spinning—they change their direction (second derivative) all the time—they require us use very small time step sizes. At the same time, they have only a small impact on the trajectories of all other heavier objects. Yet their impact on the overall system evolution is not negligible. It plays a role. Few fast objects make the overall problem stiff.

Definition 20.5 (*Local time stepping*) *Local time stepping* (LTS) techniques circumscribe that some parts of a simulation progress in time with their own (local) time step sizes.

In a multibody simulation, each particle is subject to its own set of ODEs affecting its velocity and position. The ODEs directly tell us when these properties change dramatically over one time step: The position changes quickly if the particle's velocity is high. The velocity changes quickly if the particle's mass is small compared to others or if many other particles roam nearby. With our time step adaptivity techniques from above, we can determine an optimal time step size per particle and make each particle progress in time at its own (numerical) pace. There are a few design decisions to make:

Interpolate. Assume there are two particles p_1 and p_2, and they travel at their own speed v_1 and v_2. As $v_1 \gg v_2$, we want to use a tiny time step size for particle p_1 and update p_2 infrequently. To simulate this system we can

- send the slow particle ahead, and then keep it at $p_1(t + h_1)$ until particle p_2 has caught up; or
- send the fast particle ahead. When the small time step sizes h_2 have accumulated such that they match or exceed h_1, we update the big particle, too; or
- send the slow particle off, but memorise its old position. We can now interpolate its position per time step of the fast particle.

The latter variant is the most accurate yet most expensive one, as we have to bookkeep previous solutions. In any case, we have to be careful which positions feed into F at any point in time.

Homogeneous numerical schemes. High-order schemes work best if the trajectory of a particle is smooth, i.e. does not change rapidly. If we categorise particles into (light) particles which race around and are subject to "sudden" direction changes versus (heavy) particles which are slow, then it is a natural decision to ask whether we might be well-advised to use different numerical schemes. In this context, we have to be particularly careful with the local time step sizes however:

Bucketing. If you simulate N objects in our setup, you will end up with N different admissible time step sizes. This makes the implementation tricky. A lot of codes thus discretise and bucket the local time stepping scheme:

Let h_{max}^{adm} be the biggest admissible time step size in our simulation. Bucketing groups the particles into sets: All particles with $h_{max}^{adm}/2 < h^{adm} \le h_{max}^{adm}$ belong into class A, all particles with $h_{max}^{adm}/4 < h^{adm} \le h_{max}^{adm}/2$ belong into class B, all particles with $h_{max}^{adm}/8 < h^{adm} \le h_{max}^{adm}/4$ into C, and so forth. We advance all particles in class A with $h_{max}^{adm}/2$. This is pessimistic—some of them would be allowed to use way bigger time step sizes—but it makes our orchestration simpler. Class B particles advance in time with $h_{max}^{adm}/4$. Class C with $h_{max}^{adm}/8$ and so forth. Other layouts (discretisations) of the buckets are possible.

The biggest time step used in such a scheme is also called *macro time step*, as we know that all particles will eventually end up with such a time step; though most of them as a result of many tiny steps. Another term found in literature is *subcycling*. It indicates that some particles undergo several small time steps per big step.

20.3.3   *p*-Refinement

The chapter so far fixed the order p and alters the time step sizes driven by accuracy indicators. These stem from comparisons of different runs with different hs. An alternative metric used in literature is the comparison of two schemes with different orders over the same time step size. Whenever we run two schemes with order p

versus order $p+1$, we know that the difference between them will give us an estimate for the error:

With a pth-order scheme, we have a local error term in the Taylor series which is eliminated by a subsequent $p + 1$th-order scheme. Consequently, the difference between two solutions leaves us with the leading local error of the lower-order scheme.

While it is straightforward to run two schemes in parallel, it is also expensive; in particular for Runge-Kutta schemes where we have to memorise multiple steps. A particular attractive flavour of an RK scheme thus is RFK45 (the *Runge-Kutta-Fehlberg method*). Its Butcher tableau is given by

0						
1/4	1/4					
3/8	3/32	9/32				
12/13	1932/2197	−7200/2197	7296/2197			
1	439/216	−8	3680/513	−845/4104		
1/2	−8/27	2	−3544/2565	1859/4104	−11/40	
	16/135	0	6656/12825	28561/56430	−9/50	2/55
	25/216	0	1408/2565	2197/4104	−1/5	0

The upper part of the tableau is well-known to us. It clarifies how we compute six estimates of F. The bottom part tells us how to combine these estimates into two approximations: The upper row yields a 5th order approximation, the lower row a 4th order one. One F guess is not used at all for the two final outcomes. It feeds however into subsequent steps of the RK schemes as intermediate result. This need for an additional auxiliary solution is the price to pay for the fact that we get two approximations instead of only one.

There are different applications of error estimators for higher-order methods: We know that the error should go down with h^p, and we also know that we should use a higher-order method where possible. However, we also know that this higher order quickly breaks down if the underlying solution becomes non-smooth. With a built-in error estimator, we can dynamically change the order, or we can even run codes where a time step is rolled back whenever we find out that the error did grow too much (optimistic time stepping). That is, we run RFK45. If the resulting error is above a given threshold, we roll back and rerun the time step with an explicit Euler, e.g., yet with very fine time step sizes.

20.4 Hard-Coded Refinement Strategies

The crucial element behind all of our adaptive schemes so far is a proper error estimator. Once we get an idea of the quantity of the error that we face, we can decide which time step size or numerical scheme to use. Error estimators are unfortunately not for free.

In many applications, we have a good feeling (expert knowledge) where big errors arise if we are not picky with the time step size choices: The density of particles around a particle of interest gives us a good feeling of how massive trajectories change. Small particles are, by definition, more "vulnerable" than heavier ones. If we have clusters of particles, these clusters will interact strongly with each other, while isolated particles far away only alter their flight direction slowly. More general, we know that a large right-hand side of the velocity ODE forces us to use accurate time stepping with a small time step size. Often, an analysis of the simulation state provides us with an estimate where these large right-hand sides arise, or we might know from previous experiments where we have to be careful with h and p. Such domain knowledge allows us to hard-code refinement criteria which are cheaper than their dynamic counterparts.

Further material

- E. Fehlberg, *Low-order classical Runge-Kutta formulas with step size control and their application to some heat transfer problems.* NASA Technical Report 315 (1969)
- R.J. LeVeque, *Finite Difference Methods for Ordinary and Partial Differential Equations: Steady-State and Time-dependent Problems.* Classics in Applied Mathematics. SIAM (2007)

> **Key points and lessons learned**

- Stiff problems can force explicit methods to use tiny time step sizes to remain stable.
- Techniques to choose the time step sizes adaptively ensure that we find a stable time step size automatically. For these schemes, we need an error estimator.
- We can alter the time step size or the numerical scheme or even the physical phenomena that we track (locally).
- Binning of admissible time step sizes helps us to bring the computational overhead down though the load balancing of adaptive local time stepping remains subject of active research.

The last part of the manuscript sketches two algorithmic ingredients of simulation codes that help to significantly reduce our codes' time-to-solution. Indeed, high order methods and local time stepping are considered a must in many research communities today.

Though only sketched, one of the fundamental challenges for computer scientists from the field of scientific computing is it to map the efficient numerical techniques onto efficient implementations. For some subchallenges, this is a straightforward engineering tasks. For others, the numerical concepts (seem to) contradict the efficient exploitation of modern computers. Local time stepping is difficult to load balance, optimistic time stepping is very sensitive to the cost and additional footprint of backups/rollbacks, higher-order introduces additional synchronisation points per time step, and so forth. A lot of these challenges are subject of active research in high performance scientific computing, and a lot of the state-of-the-art numerical techniques have not yet found their way into mainstream simulation codes that run on the largest machines in the world.

Students reading this text know now all they need to write a first N-body code with various interesting features. If you can realise such a simulation with local time stepping and/or a higher order method on a multicore- or GPU-cluster, you can be sure that you also understood a decent part of the underlying theory. Another option for a challenging project that combines important aspects of this manuscript in a new way is to replace the gravity force with a Lennard-Jones potential. You switch from planets attracting each other via gravity to classic molecular dynamics. The additional interesting aspect here is that you can cut off the interaction, i.e. it is sufficient to "only" study the objects around a particle to obtain valid forces. This reduces the algorithm to an $\mathcal{O}(N)$ algorithm; which poses yet another challenge to realise this efficiently.

Using the Text

A

Below is a suggestion of how to structure a course around this book plus some remarks what could be added and done beyond its content. I also add some remarks on the notation used and some useful software.

A.1 Course Organisation

The enumeration below assumes that you have 50–60 min per session.

- **Session 1: Introduction**. Introduce field (Chap. 1) to highlight why scientific computing is an important subject. There should be time to introduce the lecturer, the machines (supercomputers) to be used as well as the overall lecture concept.
- **Session 2: Motivation behind parallel programming**. Remarks on computer evolution (Chap. 2) that clarifies why we have to study parallel computing techniques. I usually make students explore the Top500 webpage (http://www.top500.org), too.
- **Session 3: Model problem and floating point errors**. I introduce the model problem that is used throughout course (Chap. 3) and start to discuss Session 4 in the remaining time.
- **Session 4: Floating point data representations**. This Chap. 4 is, for most of the students, a recap of something they should already know from computer architecture sessions.
- **Session 5: Vector computers**. A first encounter with SIMD (Chap. 5). This is a short chapter. I therefore bring content from the subsequent chapter forward:
- **Session 6: Floating point error propagation**. In Chap. 6, I discuss floating point formats and start to discuss topics around floating point precision and stability.
- **Session 7: Vectorisation**. I mix the discussion of vectorisation with a glimpse into GPGPU architectures (Chap. 7). However, the majority of this time is spent on wrapping up Session 6.

© The Editor(s) (if applicable) and The Author(s), under exclusive license to Springer Nature Switzerland AG 2021
T. Weinzierl, *Principles of Parallel Scientific Computing*, Undergraduate Topics in Computer Science, https://doi.org/10.1007/978-3-030-76194-3

- **Session 8: Stability**. We introduce stability in Chap. 8.
- **Session 9: Programming vector computers**. A glimpse into vectorisation. We start with compiler feedback and then insert pragmas. This session also comprises a high-level overview of HPC programming paradigms. Chapter 9 wraps up the number crunching part, is well-suited for some live experiments, and becomes the transition into a first lab-ish block:
- **Session 10—Intermezzo—Tools**. This is not covered by the manuscript, but it might be wise to discuss a few basic tools before you continue. Most students have never systematically been taught how to develop larger pieces of scientific software. I thus propose to have at least one lab session where students learn (or at least see)

1. how to use the simple `time` command and how compiler optimisation flags affect the outcome;
2. that compiler options have a significant impact on the runtime;
3. how to use a profiler. Starting with `gprof` is a good option;
4. how to use a debugger (`gdb`, e.g.) to find errors systematically;
5. how to run `valgrind` to identify memory leaks. An alternative is the Intel Inspector or the Clang correctness checker toolchain, e.g.

- **Session 11: Conditioning and well-posedness**. Discusses this key idea behind numerical algorithms and introduces the idea behind backward stability.
- **Session 12: Taylor and ODEs**. This is mainly a revision session. Most students will have studied these ideas throughout their A-levels already, but it is nevertheless good to revise it. I usually skip the remarks on stationary solutions, attractiveness, …and squeeze the two Chaps. 11 and 12 into one session.
- **Session 13: Stability**. I return to the attractiveness/stability definition from Chap. 12 and introduce flavours of (discretisation) stability. If you play around with the code snippets from Chap. 13, then there might not be enough time to introduce convergence and convergence plots.
- **Session 14: Convergence**. I wrap up the core block on numerics with a discussion of convergence and convergence measurements plus plots. There should be some time left to run through parts of the **Parallelisation Paradigms** from Chap. 14; as a cliff hanger.
- **Session 15: Upscaling laws and shared memory parallelisation with BSP**. This session covers the remainder of Chap. 14—I usually do a "live demo" how to determine machine parallelism from `likwid-topology` or vendor descriptions— and continues with Chap. 15. The discussion of proper upscaling plots can be left to the students if they have to provide such plots as part of their coursework anyway.
- **Session 16 and 17: BSP in OpenMP**. This session runs through Chap. 16. As I also discuss variable visibilies and collective computations as well as scheduling, I typically need two hours; notably if I want to do some live programming.
- **(skipped): Task-based parallelisation**. Chapter 17 discusses the majority of tasks. I think this is an important topic for modern codes, but my students typically prefer

to learn something about GPGPUs rather and I want to speak about something beyond explicit Euler and about proper performance analysis, too. The task chapter is thus one I often "sacrifice" and leave it as self-study exercise.

- **Session 18: GPGPUs**. Chapter 18 is popular with students. As I typically skip the task discussion and thus have to omit the fancy GPU concepts, this one fits into one session.
- **Session 19—Intermezzo—Parallel Debugging and Performance Analysis**. I propose to use at least one session to discuss tools like VTune, APS, Map or other tools. The NVIDIA analysis tools are another attractive option. The most important message behind these sessions is not to make students prolific in using the tools. The message that has to come across is that these tools do exist, that they are worth learning, and that there is tons of good material out there—most of it provided for free by the vendors. Another useful live demo session orbits around parallel debugging.
- **Session 20: Higher-order (RK) methods or adaptive time stepping**. You might want to add some more in-depth stuff on top of the content of Chaps. 19 and 20, but there is some core content on advanced algorithms in the manuscript.

A.2 What I Have Not Covered (But Maybe Should Have Done)

There are a few things that make up a natural second part of the presented material:

- I have not spoken about any linear algebra. This is one natural next topic.
- Partial differential equations are missing. I focus exclusively on ordinary differential equations, as students, in my experience, like particles that dance around, and ODEs are more intuitively accessible.
- The performance models are rudimentary.
- MPI is missing. This is the big missing thing from a high-performance computing perspective.

A.3 Notation Used Throughout the Book

Usually, I use x, y, z and t as arguments to functions such as $f(t)$ or $f(x)$. I do not distinguish vector values from scalar values, so x sometimes can be a scalar and sometimes a vector. The only exception is t, as I use t for time. It always denotes a scalar.

Whenever I work with vectors and do not use different symbols such as x, y, z, I use the subscript $_{1, 2, \ldots}$ to identify their components. In some cases, this can force me to use multiindices: If there is a vector x of positions and each entry has three com-

ponents for the three directions, then I write $x_{1,1}, x_{1,2}, x_{1,3}, x_{2,1}, x_{2,2}, x_{2,3}, x_{3,1}, \ldots$
Besides this, I try to avoid multi-indices.

My indices (loop variables in C) are usually drawn from $i, j, \ldots, n$. They are lowercase. I use uppercase letters for upper bounds, i.e. $1 \leq n \leq N$ or $0 \leq n < N$. Sometimes, I start counting with 1 in my maths formulae and use $0 \leq n < N$ in the corresponding C code. Please be aware of such index shifts.

By default, I use lowercase letters such as $f(t)$ or $y(t)$ for normal functions, and uppercase F for higher-order functions accepting further functions as arguments. Again f or F can either be scalar or vector-valued (with $f_1(x)$, $f_2(x)$, $\ldots$). It makes sense in a couple of situations to use other symbols. I use $p(x)$ for a position, e.g., and I use $v(x)$ for velocities.

C is something I use (besides the one exception with Dennard scaling in Sect. 2.2) as a generic constant, i.e. it is just a fixed positive value. If multiple Cs are found within one formula, they might have different values (but each one a distinguished fixed one). As C is just a generic identifier that says "this value is fixed, but we might not know it", any combination (additional or multiplication) of Cs yields a new constant. Besides C, I also use $a, b, c, \ldots$ or $a_1, a_2, a_3, \ldots$, e.g., as constants; in particular when they calibrate shape functions.

e is the letter I use for (absolute) error. ϵ can be the relative error, but usually I reserve it for machine precision (so, yes, a particular type of relative error). Depending on the context, it can be negative: Most books define errors as absolute values. In few cases, I however want to point out that an error might be "too much" or "too small" compared to a wanted outcome.

Small variations of entries (due to errors or normalisation, e.g.) are always denoted by Greek letters, too.

The term dt is used as difference between two t values. It means "delta t". In literature, the symbol Δt is often used instead of dt. I mainly try to stick to dt, as Δ has a different meaning in the context of partial differential equations (out of scope here). The ∂ and d symbols finally are used for derivatives. I usually use the subscript to denote what type of derivative I mean, i.e. ∂_t and d_t are the derivative in time t. Some rationale behind this as well as links to other notations are discussed on page 22.

A.4 Useful Software

Here is a list of software that I frequently use in my course. Besides the compiler, nothing is required to master the book's content, but some of these tools make students' life easier. I enlist only software that is free for students:

- You need a reasonably up-to-date C/C++ compiler. I recommend at least GCC 9.x, Intel 20.x, or LLVM/Clang 8.0. Visual Studio 2019 or newer supports the majority of features discussed, too.

- The LIKWID Performance Tools (https://hpc.fau.de/research/tools/likwid) are an extremely useful collection of Linux command line tools to measure and characterise performance.
- Valgrind/memchecker is available for almost any Linux system and helps students to detect bugs.
- Intel offers its Parallel Studio line free of charge to students. The VTune and Inspector tool shipped with the studio are, besides the core C/C++ compiler, the tools I use most frequently in class.
- NVIDIA Nsight Systems is an interesting alternative to the Intel performance tools. It also allows you to benchmark GPU codes and thus is of value for the advanced sections of the present text.

Cheat Sheet: System Benchmarking

Before we start to assess any code of ours or to run any performance studies, we need an understanding what we might be able to achieve: Modern machines have hundreds of different settings and options, and we can combine them in various ways. If we use an inferior setup, we pay in terms of time-to-solution. The other way round, we might see excellent speedup if we work with a poor baseline. But such excellent speedup is fake; it typically results from an "artificially" decreased serial ratio f.

Let's first understand what we can theoretically expect from the machine's nodes.[1] If you bike to work and want to assess you fitness level, you cannot expect to have an average speed of 100 km/h. That is the wrong calibration! It might hold for a car but not for your bike. Only the knowledge of how much we can expect allows us to make meaningful statements about the quality of our code and our setup.

The data acquisition exercise to answer our capability question consists of three steps.

B.1 Looking Up Your Hardware

The first one is trivial:

- Find out what processor you use and what its clock frequency is. On a Unix system, a simple

---

[1] For a first code run-through or tuning session, you might want to skip this section. Feel free to do so. But in the end, you will have to tell your readers how good your code performs relative to the hardware's capabilities.

T. Weinzierl, *Principles of Parallel Scientific Computing*, Undergraduate Topics in Computer Science, https://doi.org/10.1007/978-3-030-76194-3

```
cat /proc/cpuinfo
```

typically gives you all the info you need: You get the CPU make (something like
`Intel(R) Core(TM)`, e.g.) including the base frequency.

- Now go to the Internet and search for the specification of this chip. It should also
  tell you which instruction set the chip does support. Search for something alike
  AVX512, e.g. This way, you know how much you can expect from SIMDsation.

Next, we find out how many cores you have on your system. You can take this
from your supercomputer's homepage, you can extract it from the `cat` output (don't
be fooled by the hyperthreads that are displayed there), or you can use a tool like
LIKWID[2] : LIKWID gives you a set of useful tools wrapping around processor-
specific APIs, and it offers command line tools to translate these data into digestible
information. For step two, we simply call

```
likwid-topology
```

which gives us all information (and way more data) we need at the moment.

B.2   Compute the Theoretical Capability (Peak Performance)

With the number of cores and the instruction set/specification, we can compute the
theoretical number of floating point operations per second that we could get on our
machine:

$$\text{peak} = \alpha \cdot F \cdot N_{\text{Cores}} \cdot N_{\text{FPUs/Core}} \cdot N_{\text{instr/cycle}}$$

Scaling by the frequency F gives us the number of steps (cycles) our chip can
complete per second. We multiply this quantity with the number of cores on our
system times the number of floating point units. The latter is usually one or two.
Hyperthreads do not count, as they don't own FPUs of their own. Finally, we take
this number and multiply it with the number of operations our vector unit can deliver

---

[2] https://hpc.fau.de/research/tools/LIKWID/.

per cycle. For AVX512 and double precision, we have 512 bits per register. This is 64 bytes. We can hence squeeze eight double precision values into each register $(8 \cdot 8 \cdot 8 = 512)$. Plain SIMD hence yields $N_{instr/cycle} = 8$. If we can write our code with FMA, we even get $N_{instr/cycle} = 2 \cdot 8$, as we do a multiplication plus an addition per cycle. The remaining α is a little bit nasty. Processors often downclock chips in AVX mode—though it is up to the sysad to disable this feature—or clock them up (turboboost) if the power envelope (temperature) allows it. $\alpha \approx 1$ is a reasonable first guess. The final result is the magic target. It tells us how many GFlops our system can deliver; in theory. If we hit this target, then our code is clearly *compute-bound*. This is our ultimate goal—as long as we use the best numerical techniques we are aware of.

Compute Nodes Versus Login Nodes

If you run on a supercomputer, you are interested in the spec of the compute nodes, not the spec of the login nodes. You have to log into a compute node (use `salloc` for SLURM, e.g.) and run the scripts there. Many supercomputing centres make the login nodes more powerful than the compute nodes, as they are shared by multiple users. So if you benchmark this one, your data will be noisy (as other users tend to work on the login node, too) and it will characterise the wrong computer.

B.3   Determine Effective Bandwidth

In a third and final step, we acknowledge that the theoretical peak is kind of useless for many codes, as the memory will not be able to move that much data into the registers on time. To assess our code, we need a detailed understanding how much data our memory could—for a real problem, not according to some marketing presentations— move in and out of the chip. For this, we use the STREAM benchmark.[3]

"Cheating" with STREAM

If you use a STREAM benchmark code which is "unfortunately" configured, i.e. does not exploit all hardware features, you will get a too pessimistic baseline. In return, your code later might seem to run "too good". I recommend that you return to the STREAM benchmark once you have tried out the performance tuning remarks in Sect. C.1.

---

[3] https://www.cs.virginia.edu/STREAM.

STREAM runs four simple benchmarks. It

$$
\begin{aligned}
\text{copies a vector} \quad & f(x) = x, \\
\text{scales a vector} \quad & f(x) = \alpha x, \\
\text{sums up to vectors} \quad & f(x, y) = x + y, \quad \text{and} \\
\text{compuates the "triad"} \quad & f(x, y) = x + \alpha y
\end{aligned}
$$

for large vectors $x, y \in \mathbb{R}^n$, $\alpha \in \mathbb{R}$. Per benchmark, it counts how many bytes per second the memory subsystem really delivers. We know that our SIMD registers are well-suited for such operations, and hence can assume that none of these benchmarks is compute-bound. They have different memory access characteristics, and thus quantify for different access patterns how many bytes the memory subsystem can deliver per core per second to our ALUs. They are *memory-bound* and characterise the *von Neumann bottleneck* quantitatively.

Calibrated Insight

It is convenient and fair if you quantify achieved performance in time-to-solution plus percentage of peak plus percentage of STREAM throughput. These three quantities give a good picture of a code's behaviour: They tell the reader how long to wait for a result and to which degree an implementation exploits a given hardware.

⊘ Checklist

Before you start, try to get a sound understanding of your system. Here is a checklist what you might want to find out before you start your measurements or write a report. This collection is not comprehensive:

☐ Find out what your machine looks like. This should include the instruction set (vectorisation), frequency, and core counts.
☐ Summarise this insight. Do not expect a reader to look up how many cores or frequency or whatever your machine has. Do not just throw a machine name into the write-up. You can not expect the reader of your report to recherche all these data, machines change over time, or machine information might not be public.
☐ State the theoretical peak. What can we expect from your code in the best case?
☐ State the memory bandwidth. This is where STREAM data goes into. What can we expect from your code in the best case?
☐ Relate all data you present to the (theoretical) best-case quantities.
☐ …

Further Reading

- J.D. McCalpin, *Memory Bandwidth and Machine Balance in Current High Performance Computers*, IEEE Computer Society Technical Committee on Computer Architecture (TCCA) Newsletter, pp. 19–25, (1995)
- T. Röhl, J. Eitzinger, G. Hager, G. Wellein, *LIKWID Monitoring Stack: A Flexible Framework Enabling Job Specific Performance monitoring for the masses*, 2017 IEEE International Conference on Cluster Computing (CLUSTER), pp. 781–784, (2017) https://doi.org/10.1109/CLUSTER.2017.115

Cheat Sheet: Performance Assessment

C

This is my cheat sheet, i.e. my template, how to run a first performance assessment and start tuning a code. I do it top-down, I keep it generic, and I run everything as black box initially. Often, I find it useful even for my own codes to lean back and to pretend I knew nothing about them. This stops me from running into a certain direction following my own runtime hypotheses—which more often than not turn out to be wrong. Before you start to assess your code, ensure you know how your machine behaves (Appendix B).

Symbol Information

Different to a plain performance tuning, our analysis should give us some explanations why certain runtime patterns arise or where runtime flaws come from. For this, you have to give the computer the opportunity to map machine instructions (back) to source code. If you add -g, the compiler enriches the machine code with *symbol information*, i.e. helper data. Without -g, it is totally unclear where machine instructions originally come from, i.e. which high-level construct generates which machine level instructions.

C.1 Compiler Options

Let's find out which environment setup and code configuration gives us the best performance. Compilers today accept an astonishing number of parameters to guide and control the translation process. When we run our tests, we have to ensure we use the best compiler options, and that we deactivate everything that obscures our data.

- In practice, that means we use the highest compiler optimisation level (typically -O3 or -Ofast), disable all auxiliary output (all of your prints) and switch off input and output (IO).

© The Editor(s) (if applicable) and The Author(s), under exclusive license to Springer
Nature Switzerland AG 2021
T. Weinzierl, *Principles of Parallel Scientific Computing*, Undergraduate Topics
in Computer Science, https://doi.org/10.1007/978-3-030-76194-3

- On most machines, it also pays off to compile such that the compiler builds a code that's tailored towards the particular machine. With LLVW, `-march=native -mtune=native` makes the compile create the best machine instruction scheme tailored towards the particular machine you compile on. With GNU, `-march=native` has the same effect.
- Finally, compilers yield particular fast code if you use further specialised optimisation. You have to study their documentation and handbooks for your particular system.

Cross-Compilation

On many supercomputers, the login nodes are of the same make as the compute nodes. They might have more memory or more cores as they are shared amongst all users, but otherwise understand the same instruction set. If this is not the case, you have to use *cross-compilation*, i.e. you have to tell the compiler explicitly to write out machine code for a machine it is not running on.

This is the time to search the internet for details and to read the compiler manual. Last but not least, test with different compilers if you have different compilers installed. The compilers all use different heuristics and internal optimisation algorithms, i.e. they all perform differently for different test cases.

Feedback to Guide Heuristics

Compilers rely on many heuristics internally. With a feedback mechanism you can replace or tweak these heuristics: you run the code, you track certain characteristic behaviour, and then you use these lessons learned to recompile your application into more efficient code. This pattern is called *feedback optimisation*. Consult your compiler handbook for details.

C.2   Getting OpenMP Right

In principle, OpenMP runs are close to trivial. You set the environment variable `OMP_NUM_PROCS` to pick the right number of cores and run your experiment. However, there are a few things to consider.

Most supercomputer schedulers such as SLURM installations already set the right core count automatically. I do recommend that you submit a test script and add a simple `export` to it. That allows you to check once which type of variables are set by SLURM on your system. If your SLURM installation does set the core count, then I recommend not to overwrite `OMP_NUM_PROCS`. Instead, use the SLURM parameters within the script file header to configure your run. Whenever possible, leave it to SLURM to then set the right variables. Otherwise, you quickly run into data inconsistencies if you alter the SLURM parameters but forget to update your script's variables.

Fig. C.1 Schematic illustration of a two-socket system with four cores per socket, i.e. eight cores in total. To our code, there seems to be one big L3 cache and all of the cores seem to share the memory. Technically, they don't do so. The memory is split into pieces and each socket has one memory controller. If a core accesses a particular piece of memory, the actual memory access time depends on whether this memory can be reached through the local memory controller. If not, we have to hop to the other socket first

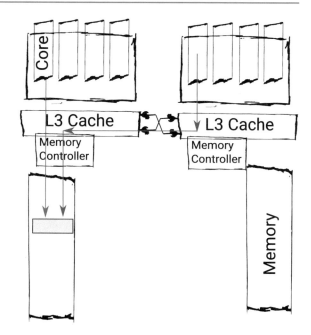

Hard-Coded Thread Count

Do not hard-code the processor count or set it manually in your code. There are very few codes which benefit from a manual specification. If your code later on has to pass scalability tests, you might even be penalised for hard-coding the thread count. Leave it to OpenMP!

The most fundamental OpenMP tuning parameter is the OMP_PROC_BIND variable. OpenMP works with threads. It is the job of the operating system to assign threads to actual hardware threads/cores, i.e. to decide where a (logical) thread is physically executed. It may even move around threads, i.e. decide to migrate them at runtime.

To move threads at runtime is typically not a good idea as this causes some overhead. You should forbid it. Furthermore, the assignment of logical threads to physical threads can be tricky: On a multi-socket system (most systems today have two sockets, i.e. half of the cores sit on one piece of metal, the other half on the others), each socket has one dedicated memory controller (see remarks on page 171). That is, each socket can access its part of the main memory particularly efficient (Fig. C.1).

To get the most out of a system, you might thus decide to spread your OpenMP threads over multiple sockets. Even if your system can keep only two threads busy, such a spread-out arrangement can keep both memory controllers busy with full efficiency. The price to pay is that shared memory exchange between these two threads might become slightly more expensive. In OpenMP

```
export OMP_PROC_BIND=spread
```

ensures that threads are spread-out. The other way round, setting the variable to `close` keeps the threads together.

Fiddling around with where threads physically go is called *thread binding* or *thread pinning*. It pays off to test different variants from time to time. If you need even more fine-granular control, you can use the environment variable `OMP_PLACES` to pin threads to particular sockets or threads. Consult your cluster/OpenMP documentation for details. You have to test which one is better.

Some compilers such as the one by Intel offer different OpenMP backends, i.e. different implementations of the scheduler, heuristics, and so forth. It is worth checking the impact of these variants.

C.3   Assessing the Code

Once we know what we could expect from the machine and once we are sure that we use a reasonable configuration, we can analyse what we obtain from our code.

C.3.1   Code Characterisation

Characterising and explaining the runtime behaviour typically requires a tool like Intel VTune or the NSight profiler by NVIDIA or a non-commercial alternative. These tools allow you to obtain some high-level overview and to answer some fundamental questions:

1. Is the work between the cores reasonably balanced, i.e. how many cores do you effectively use?
2. Does your code use all the GPUs all the time, or are they busy only for a brief period?
3. Has your code spent significant time in synchronisation, i.e. in barriers or atomics?
4. How many MFlop/s does the code or do parts of the code deliver?
5. How many bytes are read and written to and from the main memory?

I recommend that you run these tools regularly on your codes. They are simple to use nowadays, there are a lot of good tutorials, most of them are free, and it is fun to understand why your code behaves in a certain way.

Frequency

Try to assess to which degree your computer alters the frequency. There are different applications to read out the actual frequency. `likwid-perfscope` is my favourite. If a computer reduces the frequency due to heavy AVX-usage, e.g., you cannot expect the theoretical improvement of speed that you computed.

C.3.2 Diving into the Code

Once we know how our code performs and why—is it a scalability issue or are we bandwidth-bound, e.g.—we next are interested in the code parts that are responsible for this behaviour. We switch from a mere black-box performance assessment to performance analysis.

Performance Rule of Thumb

Around 80 percent of your runtime is spent in 20 percent of your code.

To find out which code parts are responsible for some behaviour, we have two options: We can instrument the code, i.e. tell the compiler to inject routines into the generated code that keep track of the runtime burnt within some code parts, or we can sample the code. Sampling-based profiling means that an external tool stops the code in certain time intervals, and takes snapshots (which function is currently active, which cores are busy, how much memory is currently used, ...) to create some statistics. Sampling is less intrusive, i.e. introduces a lower overhead, while instrumentation can yield more details.

There are many good commercial or academic tools that help you to obtain a better understanding of how your code performs. Among the most popular ones is Intel's VTune Performance Analyser. NVIDIA has its own set of tools, too. They are free for academics/students. Finally, the VI-HPS (https://www.vi-hps.org) offers a comprehensive overview incl. tutorials of free tools which are developed by academics for other academics and not driven by major chip vendors.

C.3.3 Next Steps

Once you have identified bottlenecks in your code, you might want to alter the implementation and tune it. After that, return to Sect. C.3 and rerun the assessment. Performance analysis can be time-consuming and difficult. As it is a key activity behind high-performance software engineering, there is however a lot of good tutorials and documentation available.

> **Checklist**

When you submit a code assessment, here is a checklist what might be covered by your report. This collection is not comprehensive:

☐ Summary of compile options you have used (and explanations if you have used unorthodox ones). If necessary, this one might be supported by some benchmarks.

☐ What did a high-level (black box) performance assessment tell you? Is this code bandwidth- or compute-bound? What's your high-level judgement on the simulation?

☐ What is the vectorisation and multicore performance character of your code?

☐ Where are the hotspots? Which code parts are responsible for the observed behaviour?

☐ If you have altered the code, some before versus after plots are always good.

☐ …

Cheat Sheet: Calibrating the Upscaling Models

D

Speedup laws are great to assess the quality and potential of our code. However, we seldom know the f used in both Amdahl's and Gustafson's law. Some people sit down and count operations and then come up with an f for their code. This is cumbersome and, for larger codes, almost impossible. A better option is to run experiments.

Before you do so, ensure you use the most optimised code. It makes no sense to conduct upscaling experiments with a debug version of your executable. Furthermore, I recommend to switch off any IO. Do not read or write files, and even switch off excessive terminal output. The IO subsystem of computers is notoriously slow, i.e. writing files will dominate your runtime and the real upscaling potential is hidden behind these cost.[4]

Describe Your Testbed

To allow a reader to interpret your data correctly, you have to clarify exactly what machine you use. Tell the reader the processor type, the baseline frequency, the memory, the core count and so forth. Furthermore, specify the software environment (drivers, compilers, ...) that you used. Student reports often mention the machine's name only, but this makes it hard for externals to assess the outcome—after all, you cannot expect a reader to look up the machine himself.

Many authors describe the testbed in past tense ("we used a ...") while the actual results are present tense. The rationale behind this is simple: The machine and software environment might long have been replaced by the time someone reads your outcome, but all the insights, observations and properties you state remain valid.

---

[4] It is a completely different story obviously if you want to benchmark the scalability impact of IO on a parallel run.

T. Weinzierl, *Principles of Parallel Scientific Computing*, Undergraduate Topics in Computer Science, https://doi.org/10.1007/978-3-030-76194-3

D.1 Two-Point Calibration

Strong and weak scaling are parameterised by $f \in [0, 1]$ and the time $t(1)$ we need/would need using one core. With two unknowns, we can calibrate these laws taking only two measurements: You measure $t(p_1)$ for a certain core count p_1. And you measure $t(p_2)$. For many real-world applications, none of your runs will use $p_i = 1$ as you won't be able to squeeze something meaningful onto a single core without waiting for ages. However, you can instantiate Amdahl's or Gustafson's law twice. Here's the example for Gustafson:

$$t(1) = f \cdot t(p_1) + (1 - f) \cdot t(p_1)p_1$$
$$t(1) = f \cdot t(p_2) + (1 - f) \cdot t(p_2)p_2$$

As you know $t(p_1), t(p_2), p_1$ and p_2, you have two unknowns remaining ($t(1)$ and f), so you can solve this problem. One option is to subtract the two equations:

$$0 = f \cdot \Big(t(p_1) - t(p_2) \Big) + \Big(t(p_1)p_1 - t(p_2)p_2 \Big)$$
$$- f\Big(t(p_1)p_1 - t(p_2)p_2 \Big)$$
$$\Rightarrow f = \frac{t(p_2)p_2 - t(p_1)p_1}{t(p_1) - t(p_2) + t(p_2)p_2 - t(p_1)p_1}.$$

From (15.3), we see that we are not interested in $t(1)$ for statements on the speedup. The speedup effectively normalises the runtimes by $t(1)$.

With more than two measurements—which is actually what we should do—we have to replace our simple substitution strategy with some parameter fitting.

Strong Scaling Studies

If you run strong scaling studies, you might be tempted to play around with one simulation setup only. This is dangerous: You cannot make general claims about the serial fraction of your code, e.g., if you only exposed it to one particular data set. Validate any statement with multiple, different experiments!

D.2 Runtime Normalisation

Time Step Count

We do not parallelise in time. Therefore, it is not the simulated time that we increase for weak scaling. It is the particle count! The simulation time length actually should make no difference.

The measurement span has to be long enough to get rid of noise. Many codes have a setup phase and then run many time steps. This setup phase often consists of memory allocations and data structure preparation. It thus is notoriously difficult to parallelise. You should be very careful (and document) whether you ignore this setup time or you take it into account.

Algorithms with Non-Linear Complexity

If we operate within a weak scaling mindset, the speedup formulae assume $\mathcal{O}(N)$.

In our thought experiment behind the weak scaling, we increase the problem size as we increase the compute unit count. For most algorithms this is problematic, as most algorithms' runtime is not linear in the number of unknowns. If we double the particle count in our baseline code for example, we get a four times bigger compute load.

A similar argument holds if we study an algorithm with an adaptive time step size for different setups. One setup might be "stiffer" than the other and thus yield a different time step size. So we cannot compare runtimes directly. Things become better once we follow some simple rules:

1. Present time per time step rather than raw total runtime.
2. Depending on whether you do strong or weak scaling tests, present the time per particle update or time per force computation.
3. Discuss the problem sizes explicitly.

⊳ Checklist

When you submit a scalability assessment, here is a checklist what might want to discuss in your report. This collection is not comprehensive:

☐ Exact descriptions of all problem setups come along with report.

☐ Specify the machine you are using and all settings (incl. compiler and library versions). Do not expect the reader to recherche what kind of machine you did use, how many cores it has, how fast they are, and so forth.

☐ Have you done all experiments a couple of times and long enough to avoid measurement noise, and do you quantify this noise (if significant)? The reader has to know how reliable your data are.

☐ Have you explicitly discussed the complexity of your code and normalised all data accordingly?

☐ Ensure your descriptions clarifies which application parts you assess. Are setup and IO times included or not?

☐ Identify and discuss problem sizes that lead to a strong scaling stagnation (problem becomes too small).

☐ Be fair and compare your data to a serial run that is compiled without any parallelisation feature. How much do you have to pay for the parallelisation per se?

☐ ...

Cheat Sheet: Convergence and Stability Studies

This is my cheat sheet, i.e. my template, how to run some first convergence tests with my codes, i.e. how to ensure that the code solves the correct thing with the correct accuracy or to determine this accuracy, respectively. Most tests approach the challenge in a quite generic way. This is reasonable: The steps of any stability and convergence analysis are usually very similar and problem- and algorithm-independent. It is the interpretation of these data that requires domain knowledge and a numerics background.

Any convergence analysis follows a certain pattern using key concepts from the lecture: We first ensure that we have a consistent scheme, i.e. that the solution goes in the right direction once we make the discretisation error smaller. Then, we study the code's stability. Due to the Lax Equivalence Theorem, we can then make a statement on the code's convergence behaviour and quantify it. Before we run through this sequence of tests, we have to discuss what good test cases are.

E.1 Construct Test Scenarios

Our first consideration has to orbit around proper test cases. To construct test scenarios, we have to ask "what do we want to analyse", "what setup should make our algorithm yield the expected characteristic solution", "what do we expect to see", "how do we get data to support our claim", "how many shots do we need".

Consistency of time integrators. To assess the consistency of our code, we have to ensure that we solve the right problem as the discretisation error vanishes. To make our job easier, we first should investigate whether we know an analytical solution or can construct one. If we write an ODE solver, the Dahlquist test equation is a good test. But there might be better ones. All-time classic for particles is collision tests: We construct symmetric starting conditions (arrange all particles on a sphere around

© The Editor(s) (if applicable) and The Author(s), under exclusive license to Springer Nature Switzerland AG 2021
T. Weinzierl, *Principles of Parallel Scientific Computing*, Undergraduate Topics in Computer Science, https://doi.org/10.1007/978-3-030-76194-3

the coordinate system's origin, e.g.), and thus know where exactly particles should clash.

It is important that we always track the solution at the same point in time. If we have different time step sizes (depending on object count, e.g.), we have to take this into account and alter the time step count accordingly.

With an exact function representation, we know what we want our solver to return. We now run a series of tests for different time step sizes, where we make the discretisation error smaller and smaller. If the solutions get closer and closer to the analytical solution, we conclude that we have a consistent scheme.

Consistency of physical model. Many algorithmic optimisations (in an N-body context) do not kick in for a simple (two-body) setup: Collision of different particles or adaptive time stepping, e.g., become interesting only for larger N. In this case, we should do some literature recherche what other groups have done and whether they provide quantitative data. Study exactly how they measure the data they report on and measure the same way. It is highly unlikely that you get completely different results and all the other codes have been wrong.

For more complex physics, I do recommend to brainstorm which properties should be preserved or visible to the bare eye. Often common sense helps. Is a setup symmetric and should it thus yield a symmetric outcome? Is a setup given in three dimensions but basically boils down to a one-dimensional solution which should be visible in the outcome, too? Is a setup kind of homogeneous (particles are distributed along a Cartesian mesh, e.g.) and should thus yield a homogeneous outcome?

Challenging test setups. Once we have know that our code is correct for simple setups, we should run a few setups which challenge our algorithm: I always recommend to have at least one setup that is very stiff, i.e. suffers from a large F at one point. In the N-body context, it makes sense to study setups which either host particles with extremely differing speed or are inhomogeneously distributed. You split up the domain and assign particles in one part a very high velocity. These are convection-dominated while they drift in the other part of the domain, i.e. diffuse. This stresses any time stepping algorithm, as high speed tends to require tiny time step sizes (cmp. our stiffness discussion). If you start from a very inhomogeneous particle distribution, you again stress your algorithms as some particles are subject to quite a lot of strong interactions while others are almost not affected by their neighbours.

Smoothness of Real Solution (1/2)

You cannot expect to see higher order convergence for setups where the solution itself is not smooth. If a setup yields solutions with sudden kinks, i.e. jumps in the derivatives, you cannot expect to see higher order behaviour. Still, your scheme should be consistent even though it degenerates to a lower order.

Once we have clarified what type of behaviour we want to analyse (stiffness, e.g.), we construct a setup that is as simple as possible but still runs into the phenomena of interest. This could be two particles on a collision trajectory, e.g. Write down what you would expect to happen. What stability behaviour do you expect to see, what is smoothness of the problem that you would like to see reflected in the code outcome, … Finally, test the algorithm for various input parameters and support your claims.

It is important for a fair report that you carefully document what you assume about your solution, and that you are clear where your implementation struggles or breaks down.

Convergence behaviour. To characterise the convergence behaviour, we study the behaviour of our code for a set of different time step sizes, and track the quantity of interest (velocities or positions or distances of particles, e.g.) at the end point. Before we start, it is important that we validate two things: Does our setup have solution which is sufficiently smooth, and does our setup have solution which is sufficiently interesting?

Smoothness of Real Solution (2/2)

To measure convergence, do not sample your solution at a point where it has become (close to) stationary. Higher order time integrators are better than Euler, e.g., as they add approximations of the higher order Taylor terms to the numerical solutions. If these terms that are approximated vanish for a setup, then it will be very hard to distinguish them experimentally.

The Dahlquist test equation is a poor test if you pick a large observation time stamp, as all proper numerical schemes—independent of their order—will have produced solutions close to zero. Even for schemes differing massively in order, the results will be hard to distinguish, so you cannot measure anything. A simple solution here is to pick an observation time stamp shortly after you have unleashed the simulator, such that you still face massive changes in the solution per time step.

With a table of time step sizes versus quantity of interest, we can finally compute the accuracy. If we know the exact solution, we can do this directly. If we do not know the exact solution, we cannot compute any error. The only thing we can do is to compute the differences between subsequent approximations and to extrapolate (cmp. Sect. 13.3). Hence, we need at least three measurement points, but the more we have the better.

Fig. E.1 When we compute the average of the $\|.\|_2$ norms of these points, then we obtain something alike the solid circle. We get something that illustrates a trend. If we take the $\|.\|_{max}$ of the radii, we will get a number encoding the dotted circle (fragments), i.e. we track the worst-case

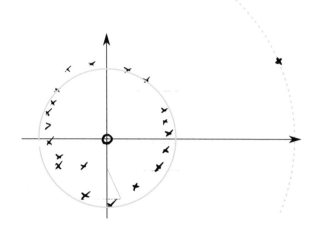

Finally, we plot the data. Log plots or log-log plots here are your friend, as we typically half the time step size per new experiment. Try to squeeze a trend line through your data points and, hence, to spot what convergence order your data has.

E.2 Choosing the Right Norms

When we plot data, we typically want to work with a scalar. We want to assign a setup a certain number. This is what a norm over a simulation state, i.e. vector, does. The two most frequently used norms are the max norm and the Eukledian norm (2-norm):

$$\|y\|_{max} = \max_i |y_i| \quad \text{and}$$

$$\|y\|_2 = \sqrt{\sum_i y_i^2}.$$

If you have N particles, you can for example report on their maximum velocity which evaluates $\|v\|_{max}$ over N quantities (remember you first have to compute the per-particle velocity via $v = \sqrt{v_1^2 + v_2^2 + v_3^2}$). Or you can report on the $\|p\|_2$ norm over the minimum distances between particles. In any case, you have to specify explicitly what you measure.

You will recognise that many authors do not write down $\|.\|_2$ but use $\|.\|_2^2$. This allows them to omit the square root.

If $\|.\|_{max}$ and $\|.\|_2$ are the popular choices, which one shall we take? This depends on the property you are interested in (Fig. E.1). The $\|.\|_2$ is an averaging norm. If all entries within a large vector have a magnitude of around seven, the norm will return 7 scaled with the square root of the size of the vector. If a few entries are really large yet the vector is large, too, then you won't see this under the $\|.\|_2$ norm. The $\|.\|_{max}$

has a different focus: It picks the biggest entry. So if you have a large vector with sevens everywhere, but one single entry is 1,000, then $\|.\|_{max}$ will return 1,000.

Choice of Norms

The $\|.\|_{max}$ is good to study outliers, the $\|.\|_2$ is good to argue about overall trends.

Norms return a positive scalar and we thus might be tempted to compare norms of different experimental runs. This can go horribly wrong. Imaging you want to argue about the $\|.\|_2$ norm (trend) of a vector which holds 1s only. If the vector has 1,000 entries, you obtain $\sqrt{1000}$. If you rerun the experiment with a bigger setup and the vector now has 2,000 entries, you obtain $\sqrt{2000}$. You can't compare them. So it might be reasonable to calibrate/normalise the norms, i.e. to use $\|.\|_x = \frac{1}{N}\|.\|_2$. The $\|.\|_{max}$ is immune to such size changes.

E.3 Interpreting Your Data

The interpretation of your data is the really tricky part and, obviously, strongly dependent on your problem. There are few generic things to take into account however:

1. Try to spot regions where your algorithm is not A-stable, i.e. where the time step size is simply to big. In practice, it is really interesting to know where these critical time steps start. For your subsequent data analysis, it is important to exclude "unstable measurements" from any computation. You can not put a trend line through data that stem from an unstable regime.
2. Try to spot whether your algorithm becomes unstable if the time steps become tiny (zero-stability). This can be laborious: If the convergence plot stalls around machine precision, then you are fine and there's nothing to worry. However, if it stalls or the error increases before you have hit machine precision, then the code's zero-stability is not great.

With these things in mind, you can start to use your observed data to calibrate the error formula. Formally, you use the ansatz

$$|e| = Ch^p$$

to come up with the problem formulation

$$(C, p) = \arg\min_{C,p} \frac{1}{2} \sum_i \left(Ch_i^p - e(h_i)\right)^2 .$$

over all measurements with h_i. Section 13.3 discusses all required techniques.

First Order

It is really hard to cook up a method that is not at least first-order accurate. So if your method exhibits a convergence order below 1 and you are operating in a stable regime, something went horribly wrong.

What happens if we had expected $p \gg 1$ and it turns out to be 1 or, in general, lower than the expected accuracy? One of the major reasons for this (besides programming errors) can be that the function that you approximate is not sufficiently smooth. If the F of an ODE jumps around suddenly, i.e. is not differentiable, then you can not expect higher-order accuracy.

⊃ Checklist

When you submit a convergence study, here is a checklist what should be covered by your report. This collection is not comprehensive:

☐ At least one problem is solved, for which we know an analytical solution. Alternatively, a problem can be solved for which others have published reference data.

☐ Exact descriptions of all problem setups come along with report.

☐ Specified which norm is used incl. a remark why these norms are appropriate to search for certain effects.

☐ Studied interesting setups under both the $\|.\|_2$ and $\|.\|_{max}$ norm. Are they qualitatively different? What does this mean?

☐ Clear identification of unstable regions with explanation what is going on.

☐ Trend lines plus quantitative data about convergence order.

☐ Discussion of possible break-downs and instabilities.

☐ Clarification of smoothness of tested problem (how many derivatives do exist and are they well-behaved, i.e. bounded).

☐ Proper formatting of all presented data (cmp. Appendix F).

☐ …

Presenting your experimental data is more than dumping it into a spreadsheet program and pressing a "create graph" button—though this might be appropriate to get a first idea of what the data looks like. This is my cheat sheet that helps you to avoid the most "popular" flaws when you present your data.

The first question to ask yourself before you start to draw a plot is whether a table or figure is the better option. Here's a rule of thumb:

- A table is better to present quantitative data. If the exact value is of relevance, then you should rather go for a table.
- A figure with a graph is better to present qualitative data. If the reader shall be able to spot a trend, a data region where something interesting happens, or shall be able to compare different measurement series, then a figure is usually the better choice.

F.1 Tables

When you create a table, try to follow some simple guidelines:

- Table captions in most styles are put above the actual table, not below.
- Ensure the caption allows the reader to understand what is displayed in the table. The reader should not be forced to search for details in the text.
- Measurements are presented top-down within a column. Each measurement goes into a row of its own. Different quantities of interest go into different columns.
- How many valid digits do you have? Is it two digits behind the decimal point or only one? Different measurements of the same quantity should all use the same precision.

T. Weinzierl, *Principles of Parallel Scientific Computing*, Undergraduate Topics
in Computer Science, https://doi.org/10.1007/978-3-030-76194-3

X-value	F(x)	Variance	Comments
0.832	17.390	0.45	Valid measurement
0.84	17.33	0.4	
0.899	17.221	0.4	Data used in further studies.
0.832	17.1456	0.4	

Fig. F.1 Example of a rather poor table with several flaws: There is no caption that tells the reader what is shown. The text is slightly too small. It should be similar to the size used throughout the script. This is a pity as there would have been enough white space to increase the font size given your current page width. The number of significant digits per measurements is not the same per column—remember that data is grouped per column, i.e. different columns can host different precisions—and the numeric measurements should be aligned right. The text is aligned left (that is fine), but the punctuation is inconsistent: One remark ends with a full stop, the other one does not. You might argue that the variance column is useless or could be presented in a nicer way

- Text within table cells is usually aligned left, but data are aligned right. As you use the same number of valid digits, the decimal point is automatically aligned.
- Use a font within the table that is roughly as big as the text font you use throughout your write-up.
- If multiple rows of a column host exactly the same data, you might be able to write it down only once and use quotation marks. Or you might be able to highlight the few entries that are interesting/different (via bold font, e.g.). If all entries are the same or only one or two differ, it might be reasonable to remove the column altogether. This is information that might better fit into the text or the caption.
- Captions do not interpret data. They describe *what* you see.

An example for a table with several flaws is given in Fig. F.1.

F.2 Graphs

When you create a figure hosting a graph, extra care is required. Before you start, clarify what type of data you have:

- Is your data continuous, discrete or is it discrete with a certain trend? If you have speedup measurements for 1, 2 and 4 cores, then you clearly have discrete data points. When you display it later on, it does not make sense to suggest that there were data for 1.5 cores. If you have however data from 1 through 256 cores and the data is reasonably smooth, then you might argue that there is a continuous trend. If you argue about different time step sizes, then your data are continuous.
- Is your data noisy or are you reasonably confident that it is pretty much spot on? If it is noisy you might have to display the standard deviation, or you might want to work with a scatter plot where the average (the trend) is displayed as a line.

Once you understand the character of your data, try to follow some simple guidelines:

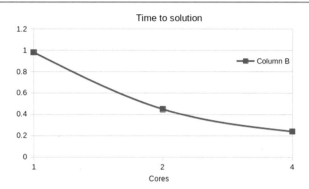

Fig. F.2 Example of a lousy figure. It is of low quality (not a vector graphics format), and one axis is not labelled. There are only three measurements—at least markers indicate where they are—but the author has decided to interpolate in-between with a smooth curve. It suggests that we have some additional information how the data behaves in-between the measurement points. The caption is above the figure, which is not in line with most journal styles, and the legend lacks a proper text

- Figure captions in most styles are put below the actual figure.
- Ensure the caption allows the reader to understand *what* is displayed in the figure. But leave the interpretation (*why*) for the text.
- Label all axes including units where appropriate. Make sure the reader does understand immediately whether you plot on a linear scale or a log scale.
- If you plot, ensure that the reader can understand the plot even if it is plotted in black and white only. It is good practice to use markers (circles, squares, triangles) to both distinguish different lines and to highlight where actual data are available and where you interpolate.
- Ensure there is a legend.
- Use a font for all labels, the legend and the caption that is roughly as big as the text font you use in your text.

Here are some further things you should not do:

- Do not interpolate if there are no continuous data.
- If you can interpolate, do not interpolate with higher order unless you are sure that this higher order interpolation is appropriate: Connect points with a line, but do not try to fit some smooth curve through your points.
- Do not use a pixel data format such as .jpg or .png. Use a proper vector format (pdf).

A few of the flaws from above are illustrated in Figs. F.2 and F.3. The list grows longer as soon as we find multiple figures within one report. Popular "flaws" are that multiple figures show data that are to be compared by the reader, but they use different ranges on the axes. Other papers present data distributed over multiple graphs but use different colour maps and different markers for similar measurements

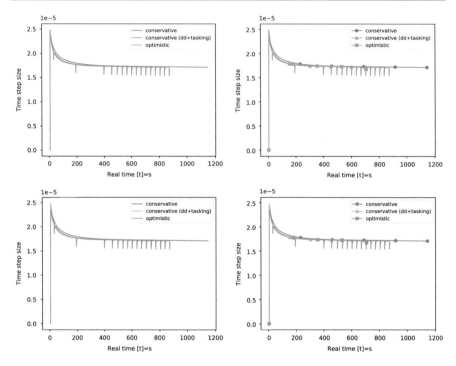

Fig. F.3 Plot without (left) and with (right) markers. The markers make the graph readable even if it is printed without colours or if a reader struggles to distinguish colours (cmp. bottom row), and they help us to argue about the three plots in an interpretation even though the graphs overlap most of the time and thus are hard to distinguish. It might be reasonable to use a symlog plot here or a log plot and to eliminate the initial 0 measurement from the plot to allow the reader to distinguish the measurements even clearer. Some authors would omit the legend from three out of four plots if they are grouped as one, while it is absolutely essential that all plots use the same scale on the x- and y-axis such that we can visually compare trends

or measurements that belong (logically) together. Make it as easy as possible for the reader:

- Keep data that belong together close (same page, e.g.) and
- use the same style for data that belongs together, such that the logical connection corresponds to a similar visual style. At the same time
- try not to present too much data per graph. Three to five data sets is a good guideline. Graphs have to be easy to digest as they communicate qualitative insight.

F.3 Formulae

Formulae are properly set by typesetting systems such as LATEX. There is not that much to do. A few pitfalls however continue to exist:

- If you give a formula a number, then you have to reference it later. If you do not refer to a particular equation later on, do not assign it a number!
- Equations are just read as part of the sentence, i.e. if I claim that

$$a \neq b,$$

  then read it out loud as "I claim that a is not equal to b. It immediately is clear that we need the comma after the formula. Formulae are part of your text and thus need proper punctuation.
- If you find a (10) in your sentence, then read it as "if you find an equation ten in your sentence". There's no need to write down "equation". The brackets do the job.
- The only exception arises if the (10) starts a sentence. In this case, you write "Equation 10 illustrates ..." and you omit the brackets.

F.4 Integrating Figures and Tables into Your Write-Up

A manuscript should be digestible without any of the figures and tables. It should be able to stand alone. However, a figure or table is considered to be non-existent, if it is not referenced within the text. When you reference it, try to be consistent: either write Figure X or Fig. X and note that most styles today consider tables and figures to be names once they come with a number. That is, Figure 10 is uppercase but "all the figures" is lowercase.

When you introduce a data set, I recommend that you first tell the reader exactly what you measure, i.e. what type of experimental setup is used. Next, describe what you see in a figure or table. This is where the reference should be found. Do not leave it to the reader to study a plot. Tell the reader exactly what you can spot. Finally, present your interpretation and evaluation. You discuss: what is done, what is seen, what does it mean; in this order.

Most type setting systems take ownership of how they move figures and tables around. However, you have some influence on this. Ensure that a figure or table is roughly placed (on the page) where it is referenced for the first time. Do not force your readers to turn pages over and over again.

Lots of style guides recommend that figures, tables and equations should not be subjects in your sentence. Indeed, I find it strange to read "Table 20 shows that the algorithm converges in $\mathcal{O}(h^2)$." Good text rather reads: "The algorithm converges in $\mathcal{O}(h^2)$ (Table 20)."

Validity

These are personal (biased) guidelines. Some supervisors will have different opinions and some journals will have other guidelines and conventions. Check them carefully!

⊘ > Checklist

When you submit a report, here is a checklist what should take into account. This collection is not comprehensive:

☐ For every single plot or table, there is a clear case why this is a table and not a figure (or the other way round).
☐ Both tables and figures are clearly labelled and captionised.
☐ All figures and tables are referenced in the text at the right place and are displayed roughly where they should be.
☐ The recommendations from the sections above either are followed or there are good reasons not to do so.
☐ All figures and tables follow the same stylesheet.
☐ You have printed everything in black and white, too, and you have checked that the reader can still distinguish all details.
☐ All captions are consistent, i.e. they either are only brief remarks or complete sentences with a full stop. Avoid inconsistencies.
☐ If multiple figures or tables use similar data sets or experiments, ensure they use the same symbols, colours, …
☐ …

Further Reading

- R.A. Day, B. Gastel, *How to Write and Publish a Scientific Paper Paperback.* Cambridge University Press (2006)
- Stephen King, *On Writing: A Memoir of the Craft.* Hodder Paperbacks (2012)
- W. Strunk Jr., The Elements of Style E.B. White, *The Elements of Style.* Allyn and Bacon Longman (1999)
- N.J. Higham, *Handbook of Writing for the Mathematical Sciences.* SIAM (2020)

Index

© The Editor(s) (if applicable) and The Author(s), under exclusive license to Springer
Nature Switzerland AG 2021
T. Weinzierl, *Principles of Parallel Scientific Computing*, Undergraduate Topics
in Computer Science, https://doi.org/10.1007/978-3-030-76194-3

Printed in the United States
by Baker & Taylor Publisher Services